Exploring
Fundamental
Particles

Exploring Fundamental Particles

Lincoln Wolfenstein
João P. Silva

CRC Press
Taylor & Francis Group
Boca Raton London New York

CRC Press is an imprint of the
Taylor & Francis Group, an **informa** business

A TAYLOR & FRANCIS BOOK

Taylor & Francis
6000 Broken Sound Parkway NW, Suite 300
Boca Raton, FL 33487-2742

© 2011 by Taylor and Francis Group, LLC
Taylor & Francis is an Informa business

No claim to original U.S. Government works

Printed in the United States of America on acid-free paper
10 9 8 7 6 5 4 3 2 1

International Standard Book Number: 978-1-4398-3612-5 (Paperback)

Library of Congress Cataloging-in-Publication Data

Wolfenstein, L. (Lincoln)
    Exploring fundamental particles / Lincoln Wolfenstein, João P. Silva.
    p. cm.
    Includes bibliographical references and index.
    ISBN 978-1-4398-3612-5 (pbk. : alk. paper)
    1. Particles (Nuclear physics) I. Silva, João Paulo. II. Title.

QC793.2.W65 2011
539.7'2--dc22   2010028439

Visit the Taylor & Francis Web site at
http://www.taylorandfrancis.com

and the CRC Press Web site at
http://www.crcpress.com

Table of Contents

Preface

A fundamental goal of science from the earliest times has been a search for the elementary constituents of the physical universe and the interactions between them. In the last half of the twentieth century, this has been the goal of a field of physics called elementary particle physics.

In the early 1930s, three particles had been identified as the constituents of matter: electrons, protons, and neutrons. Today, of these, only the electron is considered elementary, since it is believed that the protons and neutrons are made up of quarks. On the other hand, we have a list of 16 elementary particles, although most of them are not constituents of ordinary matter. In the early 1930s there were two fundamental interactions: gravity and electromagnetism. Today we have two more: the weak and strong interactions. This book is intended to explain the development of this new picture through the combined effort of theoreticians and experimentalists.

The picture that we will present in Part A of the book is called the standard model. As each aspect of it was developed, it usually took many years before it became accepted. Sometimes a new theory that became part of the standard model was ignored for several years because it had little or no experimental validation. Sometimes difficult experiments done in different laboratories gave conflicting results. New theoretical ideas often seemed too strange to believe.

The problem anyone faces in trying to explain aspects of this standard model is that, on the one hand, it should be easy to understand, but, on the other hand, the reader should realize that these ideas were only accepted after years of struggle. If you find some of the theories presented here somewhat weird, you will be in the same position as most physicists when the theories were first developed.

Although there is so much we have learned in the twentieth century, there is no reason to believe that now we have all the answers. Rather than speaking of *the* standard model, we should perhaps speak of the 2000 model of the physical universe. There may be many surprises ahead. There is a big new accelerator, the Large Hadron Collider (LHC) beginning operation in Geneva, as well as a large number of smaller important experiments in many different countries. There are many new astronomical observations being planned. In Parts B, C, and D we consider three parts of the standard model where there have been important developments in the last 20 years and which present challenges for the coming years.

In Part B, we discuss the violation of the symmetry between matter and antimatter, which goes under the name CP violation. The first small violation of this symmetry was discovered in 1964, and it took 30 years before a second different small violation was found. Only in the last dozen years have large violations of this symmetry been discovered in studies of the decay of a particle called the B meson. Experiments in progress and being planned are needed to determine whether all the violations of this symmetry are consistent with the standard model.

In Part C, experimental results involving the once-mysterious neutrino are presented. They lead to the conclusion that neutrinos have mass, which provides the first direct evidence for something beyond the standard model. New neutrino experiments are starting in Japan, China, France, and the United States. Furthermore, neutrino astronomy may offer a new window on stars and galaxies.

In Part D, we turn to the Higgs particle, the one constituent of the standard model that has not been detected. The Higgs particle plays a very central role in the standard model, and thus its detection is essential for the validation of the model. The analysis of experiments in the last 15 years has put important constraints on the mass of the Higgs particle. A major goal of the LHC accelerator is the discovery of the Higgs.

The success of human endeavors discovering fundamental new features of the physical universe is exciting and thrilling. This excitement should not be limited to a small group of scientists but should be shared with everyone. That is a major goal of this book.

Acknowledgments

The Feynman diagrams of this book were initially drawn with Freeware JaxoDraw v1.3. We are very grateful to Lukas Theussl for building a program to convert our original figures into v2.0 of JaxoDraw.

We thank our colleagues Augusto Barroso and António Silvestre, who have read the whole book and made numerous suggestions. Needless to say, ours is the blame for any inadequacies that might remain.

Stan Wakefield acted as our discerning agent, and John Navas of Taylor & Francis accepted to become our editor. We are indebted to both for their guidance and friendly advice.

Lincoln Wolfenstein wishes to acknowledge the many discussions related to this book with the late Darwin Chang, who died too young. He dedicates the book to his patient wife, Wilma, and their four children and nine grandchildren. João Silva thanks his family for their unfailing support and dedicates the book to Ana and Sara.

The Authors

The authors are professional particle physicists.

Lincoln Wolfenstein got his PhD at the University of Chicago and is professor emeritus at Carnegie Mellon University. He made landmark contributions to most of the subjects presented in this book. He is a member of the American National Academy of Sciences, and has been awarded the 1992 J. J. Sakurai Prize by the American Physical Society and the 2005 Bruno Pontecorvo Prize by the Scientific Council of the Joint Institute for Nuclear Research (Dubna, Russia). The discovery of neutrino masses, which forced a revision of the standard model, hinges on his prediction and study of the influence of matter on neutrino oscillations, now known as the Mikheyev–Smirnov–Wolfenstein effect.

João P. Silva got his PhD at Carnegie Mellon University, was a Fulbright scholar at Stanford University's SLAC, and is on the faculty at ISEL–Lisbon, Portugal. He is a coauthor of the book *CP Violation*, edited by Oxford University Press, increasingly involved in science awareness and in the interplay between science and art.

PART A

Genesis of the Standard Model

CHAPTER 1

The Foundation of Modern Physics

The Legacy of Newton

THE BEGINNING OF MODERN science most clearly is dated by the work of Isaac Newton in the second half of the seventeenth century. Much of the framework of modern physics follows from his formulations. In this introductory section we illustrate how Newton's work planted seeds that yielded many of the theories discussed in this book.

1.1 SIMPLE QUANTITATIVE LAWS

The triumph of Newton is that one can make precise quantitative predictions of observations starting from a couple of equations. Given Newton's second law and the universal laws of gravitation, one can calculate the motion of the planets around the sun, and the moon around the earth. More than that, once you see a comet come into view, you can predict its motion, and you can calculate the motion of artificial satellites that we put in orbit around the earth. All this follows from two simple equations.

To be precise, what Newton's laws tell us is that given the position and velocity of all the masses relevant to the problem, you can in principle calculate their future positions and velocities. The laws do not explain these initial conditions; they do not answer Kepler's question as to why there are six (or now eight) planets, or why they are in particular orbits around a medium-size star, our sun. The laws also allow us to extrapolate backward in time, and we can tell where the planets were a million years ago.

However, there are limits to this simple picture; we now know that the solar system is only 4.5 billion years old: we cannot extrapolate backwards indefinitely just using Newton's laws. Similarly, we now believe that in another 4.5 billion years the sun will turn into a red giant star and expand over the planets. Eventually, we must understand the internal structure of the sun and not just consider it as a massive sphere.

The calculation using Newton's laws can be done in principle, but in practice the mathematical solutions of the equations may not be easy. If we consider just one planet going around the sun, then it is easy with the calculus developed by Newton to calculate the possible elliptical orbits. When you consider several planets and include in the calculation the forces that each planet exerts on the others, in addition to the force of the sun, the calculation becomes very complicated. The problem that Adams and LeVerrier faced in determining the unknown orbit of Neptune from its effect on the motion of Uranus is even harder. But the fundamental equations that govern it all are amazingly simple.

The Newtonian world picture has sometimes been called the mechanical universe. The entire future is determined by physical laws from the initial conditions. This does not mean that in practice we can predict the future, because the more detail we want to know and the further in the future we want to see, the greater and greater the details we must know about the present. A well-known example is weather predictions. Modern quantum physics provides limits to this predictability, yet prediction from quantitative laws remains the model for physics today.

Since the time of Newton, the goal of physicists has been to discover these quantitative laws. Thus, in the nineteenth century, James Clerk Maxwell produced four differential equations that govern the phenomena of electricity and magnetism. In the twentieth, Erwin Schrödinger produced a differential equation in quantum mechanics that determines the energy states of any atom. The apparent fact that the physical world is governed by simple mathematical equations is a continual source of wonder and amazement.

1.2 FUNDAMENTAL INTERACTIONS

Newton's first law states that an object moving at a certain speed will continue to move with the same speed and in the same direction unless it is acted on by a force. If you want to speed up an object, you will have to give it a push in the direction of motion; if you want to change the direction, you have to give a push to one side or another.

In the absence of forces the universe consists of particles all moving at a constant velocity. Everything of interest that happens has to do with forces that change the motion, that may cause particles to stick together or rotate around each other.

In fact, we believe that forces act between particles; that is, the force on one particle results from the presence of other particles. Thus, the most fundamental laws of physics are the laws governing the forces between particles or fundamental interactions. This point of view continues to the present day. A major goal of elementary particle physics has been to discover the laws of interaction between particles. Today, we identify four fundamental interactions: (1) gravitational, (2) electromagnetic, (3) strong nuclear, and (4) weak nuclear.

It is the first of these force laws that was discovered by Newton. The universal law of gravitation states that every particle attracts every other particle in the universe with a force that is inversely proportional to the square of the distance between them and proportional to the product of the two masses. The proportionality constant is known as Newton's constant and is usually represented by the letter G. It is this force that determines the motion of the apple that falls from the tree and the motion of the moon around the earth.

The forces other than gravity that we are most familiar with are forces like friction or those of springs. None of these are considered fundamental forces; they depend on the details of the different materials involved. We believe that in principle, all these forces can be derived from fundamental forces acting at the atomic level, although in practice we usually use semiempirical descriptions involving parameters determined by experiment, like the Young's modulus or the coefficient of friction.

At the atomic level, by far the most important force is the electrical force. This force can be observed when combing your hair on a very dry day. You find that your hairs "become electrified" and tend to repel each other while the comb attracts your hairs. This illustrates a crucial difference between electrical forces and gravitational forces. Electrical forces can be both repulsive and attractive. In order to obtain an electrical force between two ordinary objects, it is necessary to prepare them (as by rubbing a comb against your hair) so they have a net electrical charge. There are two possibilities: positive charge or negative charge. The rule then is: like charges repel, unlike charges attract. Thus, the individual hairs with positive charge repel each other while the comb with a negative charge attracts the hairs. In 1785, by accurate measurements, Coulomb established the electrostatic law of force. For two stationary small charged objects the electrostatic force decreases with the square of the distance, just like the gravitational force. The force is proportional to the electric charge on one object multiplied by the charge on the other.

The electrostatic force law is only one component of the more general electromagnetic force laws. When the charges are moving, there is additional magnetic force. There are also magnetic forces between magnets. A complete theory of electric and magnetic forces came only with the work of James Clerk Maxwell in 1865.

On the atomic level, the electrical force is overwhelmingly larger than the gravitational force. However, on a larger scale, big objects contain about equal numbers of positive and negative charges so that electrical attraction and repulsion cancel out. Thus, on a large scale the gravitational force dominates because all particles in a large object attract other particles.

At the subatomic level, two additional interactions become important that fall off very quickly with distance. The strong force is the dominant one within the atomic nucleus and is responsible for holding the nucleus together. The weak interaction was originally formulated to explain certain radioactive decays of nuclei and now plays a very important role in the physics of elementary particles.

1.3 FIELDS

A concept that has played a major role in physics is that of the field. The simplest example is the gravitational field, for example, that of the earth. We define the field at a point in space as the force that would act on a unit mass if the mass were at that point. Thus, the field at a point outside the earth is directed toward the center of the earth with a magnitude that falls off with the square of the distance. Similarly, one can define the electric field, E, in terms of the force that would act on a unit positive charge.

At first the introduction of the field seems to add nothing to the original interaction law. However, in more complex situations it becomes essential. Given a set of moving charges Maxwell's equations allow you to calculate the electric and magnetic fields at any point. The most striking feature was that at a large distance from an oscillating charge the electric and magnetic fields varied in space like a wave, and this wave pattern moved outward with a velocity given by c, the velocity of light. These are the electromagnetic waves that vary from radio waves to light waves, to x-rays as the wavelength gets shorter.

In analyzing the photoelectric effect, Einstein pointed out that light, when it was absorbed or emitted, behaved like a particle with energy, given by hf, where f is the frequency and h is Planck's constant. Thus, there arose the wave-particle duality; one had to accept that light had both aspects. The particle is called the photon, and the probability of observing the particle at a point is proportional to the magnitude of the field.

With the development of quantum mechanics it became clear that the electron behaved like a wave, as dramatically illustrated by electron diffraction. Thus, here too there was a wave-particle duality, but the electron has a mass while the photon is massless. It became necessary to describe the electron as a field.

The fundamental equations that describe the electromagnetic interactions, called quantum electrodynamics (QED), involve electron fields and photon fields. Given the time-dependent electron field, a time-dependent photon field (or electromagnetic field) is determined, and if enough energy is available, this can correspond to the emission of a photon. On the other hand, lacking enough energy, the electromagnetic field can affect another electron; this is sometimes referred to as a virtual photon.

In the 1930s and 1940s there were a large number of successful predictions based on QED in experiments involving the emission and the annihilation of electrons and positrons as well as atomic physics. Thus, it served as the model for the development of theories of the weak and strong interactions.

1.4 COSMOLOGICAL PRINCIPLES

The story of Newton and the apple leads to the concept that the same laws of physics hold on earth and in the heavens. There is nothing special about our own time and place. We may formalize this in terms of what we will call cosmological principles:

Cosmological principle 1: The same laws of physics hold everywhere in the universe.

Cosmological principle 2: The same laws of physics hold for all times.

Perhaps it would be better to call these working hypotheses, which allow us to try to understand the astronomical universe. So far they have served us very well.

The first cosmological principle is the foundation of modern astrophysics. The laws of physics we know seem to work far beyond the solar system. We find two stars with one rotating about the other in accordance with the same law of gravity that governs our solar system. We see the same sequence of spectral lines (colors of light) coming from distant stars as those we see in our laboratory, indicating that the laws of atomic physics are the same.

Nevertheless, there is a problem of which we must be constantly aware. There may be laws of physics that we have not yet discovered. Today as we contemplate the universe, we try to apply the laws we know, but we also look at the universe as a laboratory from which we may find clues to physics not yet known.

This is wonderfully illustrated by the story of Lord Kelvin and the age of the earth and the solar system. Using the then known laws of physics, he asked for the source of the sun's energy and how long it could have been shining. If the sun were burning up by normal combustion, it could not continue more than 100,000 years. A much larger source of energy was available from gravitational collapse as the sun fell in on itself from a large size to its present radius. But even this led to a lifetime of much less than 100 million years. Such a short time seemed to contradict the theories of biological and geological evolution. The answer came only with the discovery of nuclear reactions and nuclear energy.

With the discovery of nuclear fusion reactions, it was proposed that the stellar energy was produced by nuclear reactions that fused hydrogen into helium. There was enough energy for the sun to shine for 10 billion years. Nuclear physics also brought with it natural radioactivity, which provides "clocks" that determine the age of the earth and of meteorites. A number of such clocks coincide in dating the solar system at 4.5 billion years.

The second principle is the basis of cosmology, our attempt to reconstruct the history of the universe. It has much less evidence, particularly the farther we go back in time. However, it has led to some remarkable successes over the last 40 years.

It has been proposed by some physicists that what we call physical constants are actually varying with time. Thus, Dirac had a theory in which Newton's constant, G, was not constant. Studies of the motion of the moon place a limit on the change in G of less than 1 part in 10^{10} per year, which actually disproved Dirac's specific proposal.

1.5 WAS NEWTON WRONG? THE RELATION OF NEW THEORIES TO OLD

It is often said that Einstein's theories overthrew Newton's. Was Newton wrong after all? The idea of scientific revolutions was popularized by the very interesting work of Thomas Kuhn.[1] While it correctly describes the difficulty of acceptance of new theories, it gives the incorrect impression that the old theory is to be thrown out.

The point is that, from Newton's time on, theories have been accepted only when they have been verified by empirical data. Newton's laws of motion and of gravity have described the motion of the planets and the moons and much more. How could they be wrong?

Einstein's theories of special and general relativity provide laws of motion and of gravity that look quite different from Newton's. But they do not overthrow Newton's laws; they encompass them. The new laws reduce to the old laws in appropriate limits.

Special relativity modifies the laws of motion when velocities become very large. In the limit when velocities are much less than the speed of light (3×10^8 meters per second) they reduce to Newton's laws. For ordinary motions, including those of the planets, Newton's laws will do fine. On the other hand, physicists often are concerned with electrons and other particles that are moving with speeds close to that of light.

General relativity modifies the law of gravity when the gravitational force gets very strong. In fact, it was discovered in the 1800s that there was a slight deviation of the motion of the planet Mercury, the planet closest to the sun, from the predictions of Newton's laws. (It is called the precession of the perihelion of Mercury.) No one figured out how to solve this problem, but of course they didn't say Newton was wrong; after all, his laws worked so very well and this was a small deviation. However, when Einstein developed his general theory of relativity, based on rather abstract theoretical principles, he showed that he could now explain the small problem with Mercury. For the other planets farther away, Newton will do just fine. On the other hand, we believe there exist collapsed stars and galaxy centers where gravity is so large that one must use general relativity; indeed, these may be "black holes" from which nothing can escape, no matter how fast it is moving.

Quantum mechanics becomes important at very small distances. As the distances get larger, the results become more and more the same as the results of Newton's classical mechanics. Niels Bohr called this the correspondence principle.

For new theories to be accepted today, they must encompass the old; there must be a correspondence principle.

1.6 THE ROLE OF PROBABILITY

A major distinction between Newtonian physics and physics today is that we now often talk of probabilities in contrast to the exact predictability of the mechanical world picture. There are at least three different ways in which probability enters.

1. Chaotic motion: As we have mentioned, often the outcome from an initial condition depends on the exact details. For example, if we add a planet to a given planetary system, it may stay bound or it may move away from the other planets after some time. However, even if we cannot specify the initial condition accurately enough to make a precise prediction, we may be able to give the probability of certain outcomes. A whole mathematical theory, called chaos theory, has been developed for this purpose.

2. Statistical mechanics: We often are dealing with billions of billions of particles, such as the air molecules in a room. We cannot specify all their initial positions and velocities, and we do not want to know about each individual molecule. What we can do and what we want to know is the probability that a molecule has some velocity or, on average, the fraction of the molecules that have a speed greater than some given speed. Thus, even though the motion may be deterministic, we end up talking about probabilities. This is the subject of statistical mechanics.

3. Quantum mechanics: At the atomic level the fundamental laws are not deterministic. A simple example is a radioactive atom with a half-life of a day. This means there is a 50% probability it will decay during the next 24 hours. There is no observation on it you can make that will tell when it will decay. Nevertheless, if you have a large number of atoms, you can predict with great accuracy that half of them will have decayed after 24 hours.

NOTE

1. T. S. Kuhn, *The Structure of Scientific Revolutions* (Chicago: University of Chicago Press, 1962).

CHAPTER 2

Waves That Are Particles; Particles That Are Waves

A MAJOR REVOLUTION IN OUR understanding of nature took place in the early twentieth century; we learned that light can have particle-like properties and that particles can have wave-like properties. This is deeply ingrained into the standard model of particle physics.

2.1 PARTICLES VERSUS WAVES

This book tells the exhilarating recent history of the search for the fundamental building blocks of all things and their interactions. When physicists mention "point particles," they may not be talking about fundamental particles at all. Point particles might have some internal structure, but they are so named because, whatever their internal structure might be, it has no bearing on the phenomenon under study. For example, consider a rigid ball sliding down an inclined plane without rolling and without friction. If this experiment is performed in a vacuum (that is, with all the air sucked out), the velocity that the ball has after it slides for 1 in. can be calculated ignoring what the ball is made of. It is even independent of the ball's mass; it depends exclusively on the slope of the inclined plane.

There is an interesting way to describe how this happens. When the ball is placed in a high position, we say that it has the potential to gain speed and we ascribe to it some *potential* energy. As it accelerates down the inclined plane, we say that it transforms this potential energy into *kinetic energy*, from the Greek word *kinesis*, which means motion. That is, the potential energy the ball had because it was placed in a high position is transformed into the kinetic energy associated with its speed as it moves down the plane.[1]

Another interesting quantity is the momentum of this particle. Momentum is an arrow (so-called vector) that has a size equal to the product of mass with velocity, and it has the direction of the particle's movement. Intuitively, the momentum is related to the particle's ability to push things placed in its path. Why this ability should be proportional to the velocity and also to the mass is easy to understand. Imagine that a car is sliding out of control toward you down an inclined street. If the car has a small speed as it hits you, it will push you in the direction in which it is moving and will hurt you, but you might end up okay. However, if the car has a large speed, you expect to be hurled through the air for quite a distance. It is obvious to you that the push you get goes in the direction in which the car is moving and that it increases with the car's velocity. Similarly, you expect a heavier car (say, a truck) to hurt you more than a lighter car. In accordance, momentum is proportional to mass and to velocity.

When two fundamental particles collide, the total energy in the system remains the same. That is, if you sum the energies of each particle before the collision, you get the same result as you get by summing the energies of each particle after the collision. We call this the principle of energy conservation. It is a sacrosanct law that no physicist is eager to part with. Similarly, if you sum the arrows corresponding to each particle's momentum before the collision, you get the same result as you get by summing their momenta after the collision. This is known as the principle of momentum conservation.

When you collide two particles with some internal structure, you may find that the total kinetic energy before and after the collision is the same. This is an *elastic collision*. But there are collisions for which this is not the case. Because there is conservation of energy, the only explanation is that some energy must have gone into a reorganization of the internal structure of one or even both particles. This is known as an *inelastic collision*. This idea can be used to probe the internal structure of particles. Accordingly, some experimental facilities are known as particle colliders.

We have mentioned particles. Now we turn to something completely different: waves. One can get intuition about waves by performing an experiment on a tub filled with water. Immersing a hand in the water and moving it up and down, one sees a series of water peaks and water valleys that propagate (roughly) as circles moving outward from the hand. These are the same figures one observes after throwing a stone into a pond. Imagine now moving both hands in the tub at some separation from each other. At some point, the waves produced by one hand encounter the waves produced by the other hand. What one sees next is a complicated water motion due to the fact that one wave would like to move the water surface in a certain way, while the other would like to move the water surface in a different fashion. We say that the two waves interfere.

For simplicity, assume that the two waves move in one dimension (only along a straight line), travel in opposite directions, and have the same properties. At some instant, one wave would look like Figure 2.1a and the other would look like Figure 2.1b. A water molecule situated at position A of the figure would be moved up by an amount given by the arrow of Figure 2.1a if only the first wave were present, and by an amount given by

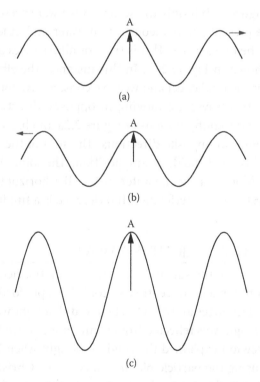

FIGURE 2.1 At a certain instant, the water molecule A is forced up by the wave in (a) moving to the right, but also by the wave in (b) moving to the left. The result is the sum of the two waves, seen in (c). This is constructive interference.

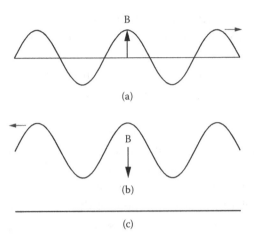

FIGURE 2.2   At a certain instant, the water molecule B is forced up by the wave in (a), but it is forced down by the wave in (b). The result is the sum of the two waves, seen in (c). This is destructive interference.

the arrow of Figure 2.1b if only the second wave were present. But, since both waves are present simultaneously, that water molecule is moved up by the sum of the two arrows. This is true for all particles along the line. The result is shown in Figure 2.1c. In this instance, the effect of the two waves enhances the final effect and we have constructive interference.

Because the two waves are moving in opposite directions, they will eventually reach the configurations of Figure 2.2a and b. Now, the arrows at position B point in opposite directions. The sum of the two arrows is zero. In fact, this result holds at any position. The joint effect is shown in Figure 2.2c. Momentarily, the water surface lies horizontally. One says that there is destructive interference. Interference is a fundamental characteristic of waves.

2.2 WHAT IS LIGHT? LIGHT IS A WAVE

Light reaches us from the sun and the stars having traveled through vast distances of empty space. For centuries people have put forth ideas on the nature of light. Descartes and Newton believed that light was a stream of particles. When light was reflected from a mirror it bounced off like a ball from a wall. Newton explained the bending of light when it passes from air into water using the particle picture. In contrast, Christian Huygens, also in the seventeenth century, developed a wave theory of light.

Around 1801, Thomas Young provided direct evidence that light behaved like a wave. He had a source of monochromatic light and shined the light onto a screen that had two very narrow parallel slits. The slits were less than a millimeter apart. Some distance behind this screen, there was another screen. On that screen he saw clearly the interference pattern consisting of alternate bright and dark lines. The fundamental difference between particles and waves is illustrated in Figures 2.3 and 2.4, following the presentation of Richard Feynman.[2]

Figure 2.3 illustrates a gun that shoots out a spray of bullets into wall A, which has two narrow vertical slits. Some of the bullets go through one of the two slits; they may go straight through or bounce off the edge of the

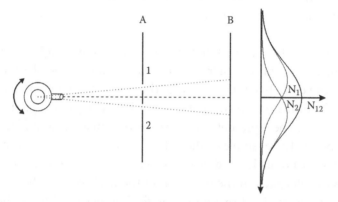

FIGURE 2.3 N_1 shows the distribution of bullets arriving at wall B when only slit 1 is open. N_2 shows the distribution when slit 2 is open. N_{12} shows the distribution when both are open.

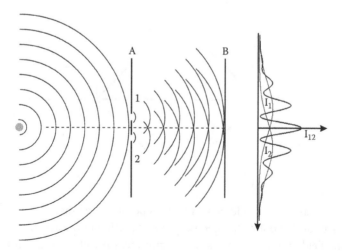

FIGURE 2.4   Young's interference experiment.

slit and change direction. If only one of the two slits is open, the resulting distribution of bullets reaching wall B is shown by the curves labeled N_1 or N_2. If both slits are open, the result is shown by N_{12}, which is simply the sum of N_1 and N_2. This is what is expected for particles.

Now consider the Young experiment with light waves. The result is shown in Figure 2.4. The intensity of the light detected at B with only slit 1 open is given by I_1. The intensity of the light detected at B with only slit 2 open is given by I_2. With both slits open, the intensity of the light detected is given by I_{12}. The crucial point is that now the intensity I_{12} is not the sum of I_1 and I_2. That is the telltale sign of interference.

In these classical experiments, waves and particles are two completely different things.

The question, then, arose: If light is a wave, what is waving? The complete answer was only attained after the work of a young James Clerk Maxwell provided the definitive formulation of the electromagnetic interaction. It requires the fundamental concept of the field discussed in Chapter 1. The electric field can be represented by a vector with magnitude and direction at every point. The example of the field due to a point charge is shown in Figure 2.5a at points located at two distances from the point charge. If we are only interested in the direction of the field, we can show the field lines that point to the direction of the field at every point, as shown in Figure 2.5b.

If we have a positive and negative charge close together forming what is called a dipole and look at the field lines some distance away, the result

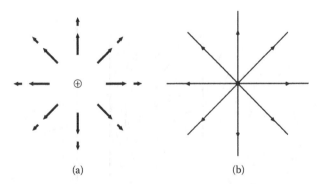

(a)   (b)

FIGURE 2.5   (a) Electric field generated by a point charge shown at points close to the charge and at points farther away. The direction of the arrow is the direction of the field and the length is the magnitude. (b) Field lines from the same source.

is shown in Figure 2.6. Since the fields due to the positive and negative charges act in opposite directions, the overall effect is reduced.

The concept of a field filling space becomes more real when it can be visualized. This was first done by the pioneering physicist Michael Faraday in the early part of the nineteenth century. He decided to place iron filings on top of a sheet of paper over a magnet; gently shaking the paper, he observed the iron filings organizing themselves along certain lines known as magnetic field lines.[3] One source for a magnetic field is a magnetic dipole like the familiar bar magnet. The magnetic field around the magnet is just like the electric field shown in Figure 2.6.

A moving charge or, equivalently, an electric current moving through a wire is another source of a magnetic field. Figure 2.7 shows the magnetic field B around a moving charge or current coming out of the page. In a few years span Faraday reproduced all experiments known to date concerning electric and magnetic effects, including Øersted's discovery that an electric current produces a magnetic field. Around 1831, Faraday sought to investigate the converse: Could magnetic fields produce electric

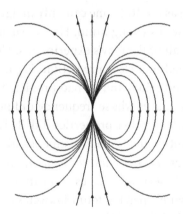

FIGURE 2.6   Field lines for the electric field generated by a dipole.

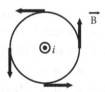

FIGURE 2.7   Magnetic field B caused by a charge or current moving out of the page. The field lines are circles centered at the current, i.

currents? He found that leaving a magnet close to an electric circuit would not induce a current in the circuit. However, moving the magnet about the circuit would induce an electric current in it. He concluded that a time variation of a magnetic field produces an electric field.

Maxwell summarized all the knowledge about electric and magnetic fields in a series of four equations. Sensing a certain symmetry between the electric and magnetic fields, he actually included a term that had no experimental verification at the time. These equations led to an amazing result when he calculated the fields at a distance from a charged particle that was rapidly oscillating up and down with a frequency f. At a point in space both the electric and magnetic fields were varying in time with the frequency f. At a given time the electric and magnetic fields varied in space just like a wave with a wavelength λ (this Greek letter is pronounced "lambda"). This whole wave pattern was moving out in space with a velocity $c = f\lambda$. The velocity was determined from Maxwell's equations.

Within the precision of the time the velocity, c, calculated by Maxwell matched exactly the known velocity for the propagation of light! Light's velocity is around 300 million meters per second (a three followed by eight zeros, usually written as 3×10^8). One can only imagine the thoughts rampaging through Maxwell's young mind. He had succeeded in providing a solid background for all of Faraday's experiments. He had invented a new term, and that term led to a clear prediction. Light should be taken as a verification of that prediction, and he had predicted that there should be other electromagnetic waves whose frequency did not correspond to that of (visible) light but whose other properties did. This can truly be ranked among one of humanity's greatest achievements, alongside Shakespeare's *Hamlet*, Picasso's *Les Demoiselles d'Avignon*, and Beethoven's Fifth Symphony. At many American colleges you can buy a T-shirt containing the words "Maxwell said:" then the four Maxwell's equations, and then the biblical "Let there be light!" Rightly so!

Sadly, Maxwell's towering achievement—the unification of electricity, magnetism, light, and yet undetected forms of electromagnetic radiation— was not fully appreciated in his lifetime. It was only 9 years after Maxwell's death that a German physicist, Heinrich Hertz, was able to produce and study electromagnetic waves. These were not light waves but more like radio waves. He used electric sparks across a small gap to produce these waves, which then propagated and produced new sparks in a suitable apparatus at the other end of his laboratory. His experiments showed that these waves had the same characteristics of visible light: reflection, refraction, polarization,

TABLE 2.1   Electromagnetic Spectrum

Types of Electromagnetic Radiation	Frequency (Hz)	Wavelength (m)
Long waves	$<10^4$	$>3 \times 10^4$
Radio waves (AM, FM, TV, etc.)	10^4 to 10^{11}	3×10^{-3} to 3×10^4
Infrared	10^{11} to 4×10^{14}	7×10^{-7} to 3×10^{-3}
Visible spectrum (red through violet)	4×10^{14} to 7×10^{14}	4×10^{-7} to 7×10^{-7}
Ultraviolet	7×10^{14} to 10^{16}	3×10^{-8} to 4×10^{-7}
X-rays	10^{16} to 10^{19}	3×10^{-11} to 3×10^{-8}
Gamma rays	$>10^{19}$	$<3 \times 10^{-11}$

Note: The frequency limits are approximate. For all waves the product of the wavelength and the frequency equals the velocity of light, 3×10^8 m/s.

and interference. Only the frequency (or wavelength) was different. Later experiments showed that these propagated with the same velocity as visible light.

We know now that there is a continuous range of frequencies for the electromagnetic waves. A rough and simplified classification is shown in Table 2.1. Notice the unit Hz used for the frequencies. This is nothing more than a unit of cycles per second, but scientists gave it the name Hertz in honor of the man who broadened our vision, in the true sense of the term. We have radio, television, and cell phones today because of the combined achievements of Faraday, Maxwell, and Hertz. Although different names are usually given to different frequency ranges, all electromagnetic waves are similar and all have the same velocity in a vacuum. The different electromagnetic interactions do not appear similar to us because we only see light (hence called visible) in the region around 4×10^{14} to 7.5×10^{14} Hz. Why do we see light in this frequency region and not in some other? Evolution! Natural selection made us "see" light in those regions of the electromagnetic spectrum for which our sun has its peak emissions.[4]

2.3 THE BIRTH OF SPECIAL RELATIVITY

For each physical wave, one may ask: What is waving? Water waves are due to the oscillation of the water surface, and the waves in a tight string are due to the oscillation of the different portions of the string. Maxwell's discovery meant that in the case of light (or any electromagnetic wave), it is the electric and magnetic fields that oscillate. The electric and magnetic fields are perpendicular to the direction of propagation of the wave and also to each other.

For a long time, scientists believed that there had to exist a medium, the so-called ether, providing the material support for light waves. In 1905 Einstein realized that no medium was required; the oscillations of the electric and magnetic fields stood by themselves, even in a vacuum. The notion of fields was here to stay. Abraham Pais[5] quotes Einstein praising "the great revolution forever linked with the names of Faraday, Maxwell, and Hertz. The lion's share in this revolution was Maxwell's. ... Since Maxwell's time, physical reality has been thought of as represented by continuous fields. ... This change in the conception of reality is the most fruitful that physics has experienced since the time of Newton."

Einstein also solved an apparent problem with the calculation of the velocity of light from Maxwell's equation. You know that different people might have different perceptions about an object's speed. For example, if you are standing still on the sidewalk and a car passes by, you can measure its velocity with respect to you. And you see the driver and the passenger inside the car move at the car's velocity. However, to the driver, both the car and the passenger are standing still. It is the outside world, including you, that the driver sees speeding backwards. Each perspective you might take is known as a *reference frame*. Once a reference frame is chosen, you measure all velocities with respect to it. But, as we saw in our example of the speeding car, the same object is tallied at different speeds with respect to different reference frames.

When calculating the speed of light from Maxwell's equations, which reference frame does it refer to? Do people in different frames measure different velocities of light? Einstein postulated that the velocity of light was *the same* in all reference frames. There was no preferred frame; there was no ether.

On September 27, 1905, Einstein published a second article on relativity, leading to physics' most famous formula: $E = mc^2$. This formula relates the mass of a particle with its energy, measured in the particle's rest frame. The significance of the rest energy becomes clear for processes in which particles disappear or are transformed, as in radioactivity. For someone who observes the particle moving, the total energy is the sum of the rest energy (mc^2) and the kinetic energy. The relation between energy, momentum, and mass in special relativity is discussed in Appendix 5. We call this the particle's energy–momentum–mass relation.

2.4 WHAT IS LIGHT? LIGHT IS A PARTICLE

The year 1905 is called the miraculous year (*annus mirabilis*) of Albert Einstein. In this year, while working full-time as a patent clerk, in his free time he wrote three papers that revolutionized our picture of the physical world. The first of these was titled "On a Heuristic Point of View about the Creation and Conversion of Light." This article seemed to imply that light behaved like a particle, despite the fact that Einstein and everyone else knew that light was an electromagnetic wave.

In 1900, Planck had developed a theory for the energy emitted by a hot object like the sun as a function of the wavelength. Technically, this is referred to as the *black-body radiation formula*. This is illustrated in Figure 2.8 for a particular temperature of the hot object; as the temperature increases, the peak of the curve moves to shorter wavelengths. You may think of this curve as the fingerprint revealing the temperature of the source.

Previous classical theories fitted the data at long wavelengths (low frequencies) but did not show the drop at short wavelengths (high frequencies). Planck found a formula that fitted the whole spectrum. To explain it he said that the radiated energy occurred in packets, each with an energy given by the product of the frequency, f, and a constant, h. For high frequencies the value of hf was larger than the average energy available in the walls, so that it became less and less probable for that energy to be emitted as hf increased. It was not clear how to interpret these energy packets, but it was a beautiful formula.

Einstein's paper discussed the photoelectric effect, the ejection of an electron from a solid surface when it is struck by light. This was first detected by Hertz in his experiments on electromagnetic waves when he observed that light striking metal produced sparks. Shortly after his discovery of

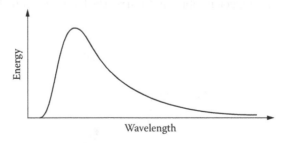

FIGURE 2.8 Black-body curve showing the radiation energy density as a function of the frequency.

the electron in 1897, J. J. Thomson showed that these sparks consisted of a stream of electrons. Further studies showed two interesting qualitative effects: (1) no matter how intense the light, there were no electrons ejected if the frequency was too low; and (2) the energy of the electron did not seem to depend on the intensity of the light but did increase with the light's frequency. Einstein explained this by saying that the light was absorbed as if it were a stream of particles, each with energy hf. There was a minimum energy W needed just to remove the electron from the solid, and so, if hf wasn't big enough, there would be no electrons. As hf increased above W, the electron energy increased and would be equal to $hf - W$. This quantitative formula for the energy was only confirmed many years later.

What does it mean that light is absorbed as if it were a stream of particles when we know light is a wave? Einstein titled his paper a *heuristic* point of view; this word might suggest an analogy rather than a reality. Only gradually was it possible to accept that light was both a wave and a particle; to explain the propagation of light, you must use the wave theory, but to explain the emission or absorption of light by atoms, you must use the particle theory. The particle picture became more accepted after the observation of the Compton effect in 1923; a photon scatters off an electron so that it looks just like the collision of two particles, as shown in Figure 2.9. Indeed, the situation can be described through the conservation of momentum and energy, treating the photon as a particle.

How can light's particle properties be consistent with Young's double-slit experiment? The answer is quite interesting. You will recall that, when comparing the behavior of particles and waves subject to the two slits, we reached the conclusions in Table 2.2. Associating photons with an electromagnetic wave gives us a conclusion from each column. Energy does come in packets (from the first entry on the "Particles" column), but the light intensity on the target is obtained correctly from the interfering waves

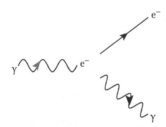

FIGURE 2.9   In the Compton scattering a photon hits an electron in a metal. As a result of the collision, they come flying off in different directions (in black).

TABLE 2.2   Classical Analysis of the Particle vs. Waves Behavior in Young's
Double-Slit Experiment

Particles	Waves
Energy comes in packets—one packet for each particle.	There is no minimum amount of energy in a wave.
After many events, the energy profile when both slits are opened is equal to the sum of the profiles obtained when one or the other slit is opened.	After many events, the energy profile when both slits are opened is *not* equal to the sum of the profiles obtained when one or the other slit is opened. This is due to interference between the waves coming from each slit.

(from the second entry on the "Waves" column). You can make this even more dramatic by reducing the intensity of the wave source in Figure 2.4 to such an extent that only one photon comes into the target at each time. You cannot predict where each individual photon will appear. Each one will bring with it a small amount of energy given exactly by Planck's constant times the source frequency. The first few photons will seem to be distributed randomly. Eventually, as the number of photons builds up, an energy profile like that labeled by I_{12} in Figure 2.4 will start to emerge.

What does it mean that light can be both a wave and a particle? Feynman in his lectures says, "They behave in a way that is like nothing that you have ever seen before. Your experience with things that you have seen before is incomplete. … I am going to tell you what nature behaves like. If you will simply admit that maybe she does behave like this, you will find her a delightful, entrancing thing."[6]

Consider again one photon emitted from the source. You cannot predict where it will appear on the screen. You can ask which slit the photon went through, but there is no answer. You could detect the photon as it went through the slit and so know which one it went through, but that would be a totally different experiment because your detection would have perturbed the light photon. You might try to perturb it very little by your detection; however, Heisenberg showed that there is a minimum amount of disturbance required to localize the photon, and so your detection will completely change the experimental result.

You want to ask the question: What was the path of the particle in going from the source to the screen? The question has no answer unless you can devise an experiment to determine the path. But Heisenberg tells you any such experiment so changes the situation that the answer you get has nothing to do with the original question.

2.5 DE BROGLIE AND SCHRÖDINGER: THE ELECTRON AS A WAVE

The electron identified by J. J. Thomson in 1887 was the first fundamental particle to be discovered. In 1913 a problem arose concerning the electron as a normal particle. Ernest Rutherford discovered that the atom contained a very small positively charged nucleus. The atom could then be pictured as a positive nucleus with negative electrons rotating about the nucleus like planets around the sun. The problem was that classical electrodynamics required that a charge on a curved path would continuously emit electromagnetic radiation, thus losing energy. As a result, one would predict the electron to crash quickly into the nucleus, in contradiction with the experiment.

To solve this problem, Niels Bohr proposed that only certain orbits were allowed for the electron, and that there was a lowest orbit that defined the ground state of the atom. This was the beginning of the quantum theory of the atom, culminating in the Schrödinger equation.

The next step occurred around 1924, when a young French physicist, Louis de Broglie, presented his thesis with the wild idea that the electron could be considered a wave. He drew an analogy between the electron and the photon in relating the particle properties of energy and momentum to the wave properties of wavelength and frequency. This provided a correct prediction for the discrete atomic energy levels of the Bohr theory.

In the beginning of this chapter, we discussed the classical differences between particles and waves, and how Young's double-slit experiment is a wonderful way to tell the difference between the two. If the electrons behave like a wave, electrons subject to two slits of appropriate dimensions should exhibit the interference effects of Figure 2.4. The experimental results are startling. Exactly like the photon, the electron also exhibits properties from both columns in Table 2.2. A very respectable particle also behaves like a wave. The specific setup of Young's experiment was eventually performed with electrons in 1961,[7] but the electron's wave-like characteristics were well established long before that by other experiments.

Waves are affected when meeting one obstacle with a size similar to their wavelength. For an electron accelerated with the help of batteries of around 100 V, the corresponding wavelength has approximately the dimensions of an atom. These were the energies present when C. J. Davisson and L. H. Germer studied electron scattering by crystals. The strong reflection at specific angles that they found could only be explained through the electron's wave-like properties. They published their result in 1927.* That same year, G. P. Thomson shot electrons with a higher energy (smaller wavelength) through a gold foil and into a target. The figures detected on the target were characteristic of waves (a so-called diffraction pattern). For their experiments, Thomson and Davisson shared the 1937 Nobel Prize. It is quite interesting that J. J. Thomson received the Nobel Prize for showing that the electron is a particle, while his son, G. P. Thomson, received it for proving that the electron is a wave. They were both right!

When de Broglie proposed that every particle should have an associated wave, the exact nature of such a wave was not known. One could not even describe such a wave mathematically. This was in sharp contrast with electromagnetic waves. The electromagnetic wave equation had been found by Maxwell, describing oscillating electric and magnetic fields that (when squared) are related to the light intensity. In turn, light intensity is related to energy. For a given frequency, the corresponding photons have a specific energy. Thus, the waves' intensity is proportional to the number of photons. We made use of this fact at the end of the previous section.

Inspiration from electromagnetic waves led to the next two developments. In 1925 Schrödinger was able to find an equation describing the pattern of electron waves inside atoms. The wave described by Schrödinger's equation became known as the *wave function*. Its physical significance would be given by Max Born (not Niels Bohr) in 1926. In general terms, the square of the value of the wave function at some position gives the probability of finding the particle around that position.[†] This marked the start of the description of quantum particles: quantum mechanics.

The fact that we talk about probabilities in quantum mechanics has suggested to some people that we cannot make precise predictions using quantum theory. This is very misleading. When we deal with large numbers of particles or atoms we make very precise predictions of the results of any experiment. For example, because the electron has a charge and a spin (discussed in Chapter 3), it behaves like a little magnet with a strength given by the magnetic dipole moment. This has been measured

---

* The first results of Davisson were reported in 1924, around the same time as de Broglie presented his thesis; definitive results were presented in 1927.

† To be precise, the wave function is complex. It is its absolute value that gives us the probability density that a particle might be found in a particular position at a particular time.

experimentally with amazing precision. In appropriate units, it is measured to be 1.0011596521859 with an error of ±0.0000000000038. Notice that the error affects only the twelfth and thirteenth decimal places. (We can only wish that bank interest rates were calculated to this precision.) Now comes the shocker: the theoretical prediction for this quantity, which involves very deep, detailed, and difficult consequences of quantum mechanics, agrees with the experimental result within the precision mentioned. The statement that all quantum mechanical—and, by implication, all physics—knowledge is probabilistic is wrong.

NOTES

1. The kinetic energy is given by $K = mv^2/2$ and the potential energy by $U = mgh$. Here, m is the particle's mass, v its velocity, h the vertical position of the particle with respect to some chosen reference height, and g the downward acceleration all objects experience because they are subject to the earth's gravitational pull. Roughly, $g = 9.8$ ms^{-2}. In the absence of friction and air resistance, the total energy, $K + U$, is a constant. That is,

$$\tfrac{1}{2}mv^2 + mgh$$

is a constant. So, as the ball goes down the inclined plane, h gets smaller. If the sum is to remain constant, the only possibility is that simultaneously the velocity gets larger.

2. R. Feynman, R. B. Leighton, and M. L. Sands, *The Feynman Lectures on Physics* (Reading, MA: Addison-Wesley Publishing Co., 1963).

3. When an electric charge is moving, it is subject to a magnetic force. There is a magnetic field B defined at every point by the force on a moving charge. The force on a unit charge has a magnitude equal $B \times v$, where v is the speed of the moving charge. The direction of the force is perpendicular to the direction of the magnetic field and to the direction of motion of the charge. The analogue of a point electric charge is called a monopole; as far as we know, no such particle exists, although there have been many searches for one.

4. When in a vacuum, all electromagnetic waves satisfy $c = \lambda f$, where c is the velocity of light in vacuum, f is the frequency, and λ is the wavelength. If you take an instantaneous picture of the wave, its wavelength is the distance between two crests, which coincides with the distance between two valleys. Since the various electromagnetic radiations differ by their frequency and all have the same velocity, they also differ in their wavelength. For violet light,

$$\lambda = \frac{c}{f} = \frac{3 \times 10^8 \, \text{m/s}}{7.5 \times 10^{14} \, /\text{s}} = 400 \times 10^{-9} \, \text{m}$$

The human eye can detect radiation from around 400 trillionth of a meter (violet: 400×10^{-9} m) until around 700 trillionth of a meter (red: 700×10^{-9} m). For comparison, the size of the carbon atom is around ten thousand times smaller than red light's wavelength and comparable to the wavelength of x-rays.

5. A. Pais, *Subtle Is the Lord: The Science and the Life of Albert Einstein* (Oxford: Oxford University Press, 1982), p. 319.
6. R. Feynman, *The Character of Physical Law* (Reading, MA: Addison-Wesley Publishing Co., 1963).
7. C. Jönsson, *Z. Phys.*, 161, 454 (1961).

CHAPTER 3

Particles That Spin

3.1 ANGULAR MOMENTUM AND SPIN

We have mentioned before two very important conservation laws: conservation of energy and conservation of momentum. Now we wish to tell you about another conservation law: the conservation of angular momentum.

Imagine an ice skater starting to spin while standing upright at a particular spot of the ice rink (this is known in figure skating as the scratch spin or upright spin). Initially the arms are extended. To spin faster, the skater brings the arms closer to the body. A body's inertia to rotation is smaller if its mass is distributed closer to the axis around which the body spins. Thus, when the skater brings the arms closer to the body, his or her inertia to rotation decreases. We define the *angular momentum* as the product of the inertia to rotation and the spin velocity. The *spin velocity* is proportional to how many times in a minute the skater goes completely around. Note that this is analogous to the definition of momentum, which equals the mass (which measures inertia) times the linear velocity. In the situation described, the skater's angular momentum is constant: as the skater brings the arms closer to the body, the decrease in the inertia is compensated by an increase in spin velocity.[1] This is exactly what we observe.

Another instance where the angular momentum conservation plays an interesting role concerns the motion of the earth and the other planets around the sun. You know that the planets move around the sun. But they do not move in circles around the sun; rather, they describe a slightly squashed circle known as an ellipse, with the sun off center.

In 1609, two events took place that would revolutionize our view of the cosmos. Galileo produced the most powerful telescope of his time, which he turned to the moon, Venus, and Jupiter. And Kepler published his *Astronomia Nova*, where he used Tycho Brahe's meticulous astronomical observations to show that the planets move in ellipses, not circles. In this motion, the angular momentum of the planet with respect to the sun is constant. As a result, when the planet is closer to the sun, its velocity should increase; when the planet is farther from the sun, its velocity should decrease. This is what Kepler deduced from his analysis of the planet's motions, and it is codified in Kepler's second law. Of course, Kepler deduced this from observations; it was only much later that it was realized that this was an illustration of the conservation of angular momentum.

The angular momentum is defined not only by a magnitude, but also by a direction—an arrow. The direction is defined as perpendicular to the plane in which the rotation occurs; thus, for the skater on ice it is perpendicular to the ice. For a given plane there are two possibilities, depending on whether the rotation is clockwise or counterclockwise. This is illustrated by the grey arrow in Figure 3.1. Imagine that you are standing in front of the plane where the motion is taking place. If you see the particle moving clockwise, the angular momentum is directed away from you, as in Figure 3.1a. If you see the particle moving

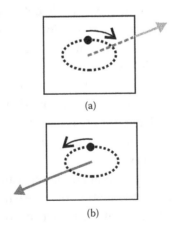

(a)

(b)

FIGURE 3.1 The curved arrow indicates the direction in which the particle orbits. The straight grey arrow indicates the direction of the corresponding angular momentum, which is perpendicular to the plane shown: (a) into the page; (b) out of the page.

counterclockwise, the angular momentum is pointing toward you, as in Figure 3.1b.

In the Bohr theory of the atom only certain orbits were allowed. The allowed orbits were those that had an integral value (1, 2, 3, …) for the angular momentum, in units of a fundamental measure of angular momentum known as $\hbar$ and pronounced "h-bar."* Thus, in quantum mechanics the angular momentum is also quantized.

The light emitted by atoms occurs when the electrons jump from one orbit to another. The energy, E, of the photon (or the frequency of the light, $f = E/h$) is determined from the energy lost by the electron. An electron moving around the nucleus behaves like a little magnet whose strength is proportional to the angular momentum of the electron orbiting the nucleus. Therefore, in a magnetic field the energy of the photon for a given transition is changed. This was first observed by Pieter Zeeman in 1896 and is called the Zeeman effect.

As this effect was studied over many years, there were cases in which the change in the photon energy due to the magnetic field did not agree with the atomic theory. The solution of this problem was proposed by two Dutch graduate students, George Uhlenbeck and Samuel Goudsmit, in 1925. They suggested that electrons carry an angular momentum quantum number. It is as if the electron were a small sphere rotating about itself while orbiting the nucleus. This is analogous to the earth rotating about itself (thus defining the day) while it orbits the sun (which defines the year).† They postulated that this angular momentum should not be an integer times $\hbar$ but, rather, equal to $\hbar/2$. When measured along a specific axis (say, the z axis), this spin could only appear as spin up ($+\hbar/2$) or spin down ($-\hbar/2$).

This intrinsic angular momentum is an attribute of all fundamental particles. Because of the analogy with the angular momentum that determines the motion of the spinning skater, it is known as spin. One can view it as some sort of internal rotation, as illustrated in Figure 3.2. The direction of internal rotation is related to the direction of the spin by the same right-hand rule explained above. One must bear in mind that this internal rotation is just a mental picture; physical rotation does not make sense for fundamental point-like particles.

---

* Is equal to Planck's constant h divided by 2π. Note that in Bohr's theory the orbits farther from the nucleus had more angular momentum.
† The young American physicist Kronig had proposed a similar idea. The idea was ridiculed by Pauli, a famous and opinionated physicist, leading Kronig to drop it. Trust yourself!

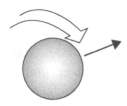

FIGURE 3.2   Relation between the imagined internal rotation, indicated by the curved arrow, and the direction for the spin of the particle, indicated by the straight grey arrow.

In units of $\hbar$ (i.e., dropping $\hbar$) all particles have an integer or a half-integer spin. In these units, the electron has spin 1/2. Particles with a half-integer spin are known as fermions, in honor of Enrico Fermi. Particles with an integer spin (including zero) are known as bosons, in honor of the Indian physicist Satyendra Nath Bose. The combination of spin and orbital angular momentum gives the total angular momentum, which is conserved in any reaction involving protons, neutrons, electrons, photons, etc.

As we shall see later, all particles of matter (quarks and leptons) have the same spin as the electron, spin 1/2. In contrast, particles like the photon that are associated with the fundamental interactions have spin 1. These include the gluons (associated with the strong interaction, as discussed in Chapter 6) and the W and Z bosons (associated with the weak interactions, discussed in Chapter 8). The almighty Higgs particle, to be discussed later, has no spin; we say its spin is zero.

3.2 HELICITY

Helicity is the projection of the spin along the direction of motion. When measured along the direction of motion, the spin of a spin 1/2 particle, such as the electron, can only take two possible values: +1/2, as in Figure 3.3a, or −1/2, as in Figure 3.3b. In Figure 3.3a the spin (in grey) points along the direction of motion (in black). The particle is said to have positive helicity. In Figure 3.3b the spin (in grey) points opposite to the direction of motion (in black). The particle is said to have negative helicity.* We stress that these figures are only meant as visual aids for fundamental particles. Fundamental particles have no dimension, and thus cannot rotate about themselves.

---

* Most physicists refer to particles of positive helicity as right-handed and to particles of negative helicity as left-handed. This creates confusion with the notion of left-handed chirality and right-handed chirality, to be introduced below.

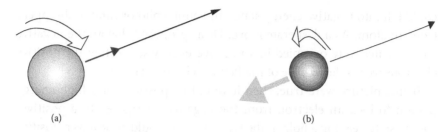

FIGURE 3.3 A spin 1/2 particle has two possible helicities. The spins appear in grey and the direction of motion in solid black. The particle in (a) is said to have positive helicity. The particle in (b) is said to have negative helicity.

There is one subtlety concerning helicities. Imagine that you have an electron. It has some mass and some velocity when measured in the laboratory reference frame. Let us assume that it has positive helicity. Now consider a reference frame that is moving in the same direction as the electron, but faster. In that reference frame, the electron is seen to be moving backward. But its internal rotation remains the same. As a result, the electron's helicity is negative in the new frame. Thus, helicity is a frame-dependent quantity. The frame dependence of the helicity would not occur if the electron had zero mass. In that case, the electron would travel at the velocity of light and no frame could travel faster than it. The electron's helicity would be the same in all frames.

As discussed later, the weak interaction has the property that a zero-mass fermion is created with negative helicity. For a massive fermion the result is that the helicity is primarily negative when the momentum is much larger than the mass. It can, however, appear as a positive helicity fermion with an amplitude proportional to its mass divided by its momentum. We refer to a fermion produced in this way as a left-handed chiral fermion.

3.3 THE DIRAC ELECTRON

In 1928 Dirac provided a formalism for describing the spinning electron consistent with Einstein's special relativity. The spinning electron behaves like a little magnet because it can be considered a spinning charge. The strength of the magnet, called the magnetic moment, appeared to be twice as large as suggested by a classical analogy. Dirac's theory correctly predicted the moment.

There was a major problem, however. Dirac's equation required states of negative energy as well as positive ones. This would mean that electrons could fall into negative energy states and emit a photon, and so disappear from an atom. As a desperate move, Dirac proposed that all the negative energy states were occupied in a negative-energy sea so that the positive electrons were safe because of the Pauli exclusion principle.

If this picture were true, then it would be possible for a high-energy photon to kick an electron from the negative-energy sea to a positive-energy state, leaving a hole in the sea. The hole would appear as a positive electron, e^+, so that the result would appear as the creation of an e^+-e^- pair by the photon. In 1932 Carl Anderson discovered the positively charged electron, the positron, while studying cosmic rays.

After a number of years it was realized that you could modify the formalism to eliminate the negative-energy sea and treat the electron and positron symmetrically. Spin 1/2 particles had four states: particles with two spin states and antiparticles with two spin states. There was one possible exception. There could be a massless particle with only one spin state, say, negative helicity, and an antiparticle with positive helicity. This was called a Weyl particle. For many years it was believed that the neutrino was a Weyl particle.

3.4 POLARIZATION AND THE PHOTON SPIN

To illustrate polarization consider the string in Figure 3.4, which is free at one end, passes through a wood panel with a vertical slit, and is tied at the other end. A vertical jolt is given to the string's free end. It creates a disturbance, a pulse, which propagates along the string. Because the jolt was vertical, the pulse is also vertical. When the pulse reaches the wood panel, it forces the string to move up and down. The pulse passes unimpeded through the wood panel because the slit is also vertical.

In contrast, a horizontal jolt will create a horizontal pulse, which will not pass through the vertical slit because the string is clamped by the panel in this direction. This is shown in Figure 3.5. In both cases, the pulse is

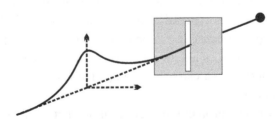

FIGURE 3.4   A vertical pulse passes through a vertical slit.

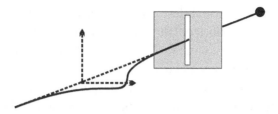

FIGURE 3.5   A horizontal pulse does not pass through a vertical slit.

perpendicular to the direction of propagation. The pulses correspond to a vertical or a horizontal motion of the string, while the propagation is longitudinal. Waves with these characteristics are named transverse waves. And the two possibilities for the pulses are known as the wave's polarization. The wave in Figure 3.4 is vertically polarized; the wave in Figure 3.5 is horizontally polarized.

We have mentioned that the electromagnetic waves in vacuum, predicted by Maxwell and confirmed experimentally by Hertz, do not correspond to motion of a physical object (as they do in the case of waves on a string), but rather to oscillations of the electric and magnetic fields. But, like waves on strings, the electromagnetic waves are also transverse. The oscillation of the electric field is always perpendicular to the direction of motion. If the electric field is directed and oscillates along the vertical direction, we say that light is vertically polarized.* If the electric field oscillates along the horizontal direction, we say that light is horizontally polarized.

Polarized light is very easy to produce. Imagine sunlight that strikes a horizontal surface at a very shallow angle. The component of the electric field that is horizontally polarized tends to move the electrons of the material along the surface. This is easy to do and most of that light is reflected. In contrast, the component of the electric field that is vertically polarized tends to move the electrons out of the material, perpendicularly to its surface. This is difficult to do and most of that light is absorbed. As a result, light reflected from a horizontal surface at shallow angles is almost all horizontally polarized.

Polaroid lenses have the same effect as the slits on the wood panels in Figures 3.4 and 3.5. The panel in those figures can serve as an analogy to

---

* The magnetic field is also transverse and, moreover, it is perpendicular to the electric field. Consider a light wave that propagates longitudinally as in Figures 3.4 and 3.5. If the electric field is directed and oscillates along the vertical, the magnetic field will be directed and oscillate along the horizontal direction, and vice versa. By convention, the polarization refers to the direction of the electric field.

polarized sunglasses, which let vertically polarized light through, while absorbing most of the horizontally polarized light. Glare originating from light's reflection on horizontal surfaces is almost all horizontally polarized and, thus, it is absorbed by polarized sunglasses.

When we go from the wave picture to the particle picture for light there is a fascinating relation between the particle spin and polarization. The particle, the photon, has spin 1. With the direction of motion as the axis, the spin can be directed parallel to the axis (+1) or opposite (–1); these correspond exactly to the two types of circular polarization known for light. Suitable linear combinations of these two correspond to the two transverse polarizations discussed above. Not all waves are transverse. Waves where the motion is along the direction in which the wave is propagating are known as longitudinal waves. For example, sound waves correspond to small oscillations of the air molecules along the direction of propagation; they are longitudinal waves.

Massive spin 1 particles, like the W and Z bosons discussed in Chapter 8, must have three possible polarizations; this is required by the principles of special relativity. With the direction of motion as the axis, the spin projection on the axis can be 0 as well as +1 or –1. In the wave picture this means there are both transverse and longitudinal waves. Conversely, this means that since electromagnetic waves are transverse, the particle picture must involve a zero-mass particle: the photon.

NOTE

1. Technically, for the rotations described here, the inertia to rotation is known as the moment of inertia and represented by the letter I. For a point-like particle of mass, m, moving in a circle of radius, R, the moment of inertia with respect to the center of the circle is given by $I = mR^2$. The notion of spin velocity is not really a velocity. It is the rate of change of the turn angle with time, also known as angular velocity and usually represented by the letter ω (omega). The angular momentum is given by the product $L = I\omega$. If a particle moves closer to the axis (R is smaller), then its moment of inertia, I, is smaller. The only way to keep the angular momentum, L, constant is for the angular velocity, ω, to increase. Thus, when a skater spins with the arms extended the spin velocity is small. By bringing the arms closer to the body (closer to the vertical axis around which the body spins) the moment of inertia decreases and the spin velocity increases.

CHAPTER 4

Understanding Quantum Electrodynamics

Feynman to the Rescue

W<small>E HAVE SEEN THAT</small> electrons and photons have both particle-like and wavelike properties. They are neither particles nor waves and should better be regarded as *quanta*. Over time, physicists recognized that reconciling quantum mechanics with relativity would require the possibility that particles be created and annihilated. A theory was needed to join all these effects: particle properties, wave properties, and particle creation.

The mathematical framework used in describing quanta is known as *quantum field theory*, and it is the consistent theory joining quantum mechanics and special relativity. We associate a quantum field with each quantum species. So, we have one quantum field for the electron and another for the photon. Each quantum field describes the respective quanta and how these quanta propagate through space-time. Quantum field theory is arguably one of the most important conceptual achievements of twentieth-century physics.

Initially, quantum fields appeared in connection with the electromagnetic field, with the electron treated as a particle. In 1929 Heisenberg and Pauli introduced the electron field. Almost two decades later, Feynman, Schwinger, and Tomonaga would develop the consistent quantum field theory of the electron and photon fields we still use today. It is known as quantum electrodynamics (QED).

4.1 QUANTUM FIELD THEORY AND FEYNMAN DIAGRAMS

In QED, the fundamental interaction is that between electrons and photons. While the theory is originally presented in the form of mathematical formulas, a simple way to understand it is with the aid of diagrams introduced by Feynman, as in Figure 4.1a, where an electron comes in and an electron and photon emerge. The point at which the three fields meet is called the vertex. The strength of the interaction is given by the electric charge, e. The probability of the process is proportional to the square of e. The processes shown in Figure 4.1 cannot occur for free incoming and outgoing particles, since that would violate the conservation of energy and momentum.* They can occur for electrons in atoms or unbound electrons moving past the nucleus of an atom. Thus, Figure 4.1a could correspond to an electron in an excited state of an atom emitting a photon and going to a lower state. It also corresponds to the emission of a photon by a fast electron passing an atom, a process known as bremsstrahlung. The rules of QED state that the same interaction can also involve a photon coming in, as in Figure 4.1b. An example would be the photoelectric effect, discussed in Einstein's original paper, in which an incident photon kicks an electron out of a metal. In all these and subsequent Feynman diagrams particles travel from left to right.

Any interaction that involves the disappearance of an electron could instead involve the creation of a positron. As mentioned in Chapter 3, in Dirac's original theory the disappearance of an electron from the negative-energy sea could be interpreted as the creation of a positron. In QED the negative-energy sea has been eliminated, but there remains the possibility of positron creation. Figure 4.2a can be obtained from Figure 4.1a by replacing the incoming electron by an outgoing positron. The resulting diagram is referred to as pair creation. Similarly, instead of the emergence of an electron and positron pair in the final state, there can be the disappearance of one. This gives Figure 4.2b, referred to as pair annihilation. The processes in Figure 4.2 are also impossible when all the particles are free.

For the case of free incoming and free outgoing particles, it is necessary to use the interaction more than once. The way to do this is made clear by Feynman diagrams. Schwinger is reported to have complained that this was "bringing computation to the masses." We couldn't agree more, but we take as laudatory what he meant as derogatory.

---

* One way to see this is to note that the process should be possible in any reference frame. If you choose the frame in which the original electron is at rest, it is obvious that the process in this frame violates the conservation of energy.

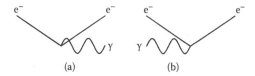

FIGURE 4.1 (a) Feynman diagram for the interaction in which an electron emits a photon. (b) Diagram for the absorption of a photon by an electron. All diagrams are to be read from left to right.

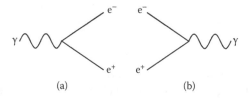

FIGURE 4.2 (a) Feynman diagram for the creation of an electron and a positron by an incoming photon. (b) Incoming electron and positron disappear and a photon comes out.

Examples in which the interaction enters twice are shown in Figure 4.3. These diagrams are consistent with the conservation of energy and momentum and the requirements that the external legs represent real physical particles.

The beauty of Feynman diagrams is twofold. On the one hand, they are descriptive of the physical process under study. On the other hand, the Feynman diagrams are related to specific mathematical rules that physicists learn on their first course on quantum field theory. For physicists, Feynman diagrams are also an extraordinary bookkeeping device allowing complicated calculations to be organized and carried out with much less effort. Calculations based on Feynman diagrams yield what is known as the amplitude for a given process. The probability, or rate, for that process is obtained by taking the square of the amplitude.[1]

Let us look at some diagrams more closely. Figure 4.3a describes the interaction of a photon with an electron. This represents the quantum field theory calculation of the process studied by Compton in 1923. It was Compton who first observed this process and applied the conservation laws to it, providing very important evidence for the photon picture.

Figure 4.3b describes the interaction between two electrons. The photon's journey between the two electrons is responsible for the

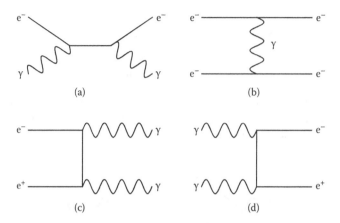

FIGURE 4.3   Feynman diagrams representing second-order interaction effects: (a) Compton scattering: $e^- + \gamma \rightarrow e^- + \gamma$; (b) electron-electron scattering: $e^- + e^- \rightarrow e^- + e^-$; (c) electron/positron annihilation: $e^- + e^+ \rightarrow \gamma + \gamma$; (d) electron/positron creation: $\gamma + \gamma \rightarrow e^- + e^+$. These diagrams are collectively known as the tree-level diagrams.

electromagnetic repulsion they feel. This can be described in many ways. We can say that the photon carries the message "there is another electron around." Accordingly, the photon is known as the carrier or the messenger particle for the electromagnetic interaction. An often used analogy describes two skaters exchanging a large ball. When throwing the ball forward, the first skater recoils. And, upon receiving the ball, the second skater gets pushed back. Of course, the analogy refers only to the repulsion in electron-electron scattering and not the attraction in electron-positron scattering. The latter process, whose diagrams we have not drawn, is one of the things that can happen when an electron and positron interact: $e^- + e^+ \rightarrow e^- + e^+$. The other possibility is shown in Figure 4.3c: the electron and positron disappear completely (annihilate) and two photons appear instead. Figure 4.3d describes the reverse process: two photons disappear and an electron-positron pair appears; you get particles where you once had light!

We mentioned that the vertex diagrams of Figures 4.1 and 4.2 cannot correspond to physical processes. This was the result of an incompatibility between the following simultaneous requirements:

1. The initial electron obeys the energy–momentum–mass relation.

2. The final electron obeys the energy–momentum–mass relation.

3. The final photon obeys the energy–momentum–mass relation.

4. There is momentum conservation at the vertex. That is, the momentum of the initial electron equals the sum of the momenta of the electron and photon in the final state.

5. There is energy conservation at the vertex. That is, the energy of the initial electron equals the sum of the energies of the electron and photon in the final state.

Thus, the vertex diagrams of Figures 4.1 and 4.2 cannot correspond to physical processes. But the same vertex appears, for example, in Figure 4.3b describing the interaction between two electrons. And this interaction does exist. How can this be? The answer must be that one of our requirements, 1 through 5, is abandoned. Clearly a real electron in the initial or final state must obey special relativity. So, we keep 1 and 2. Also, if we take the diagrammatic description of the interactions seriously, we are bound to keep also the momentum and energy conservation at the vertex: 4 and 5. Thus, the vertices can only appear in Figure 4.3b because the photon does not obey its energy–momentum–mass relation. That is, for the intermediate photon, the energy is not equal to the product of its momentum by the velocity of light. Because this relation is not satisfied, we say that the photon is off mass shell. And we can only accept this because the photon appears as an intermediate state that is not measured: we say that the photon is a virtual particle. Similarly, the intermediate electron in Figure 4.3a is also a virtual particle. We could consider Figure 4.3b as providing a quantum interpretation of the classical Coulomb interaction when the electric field is thought of as a virtual photon. Similarly, the intermediate electron in Figure 4.3a, c, and d is a virtual particle.

The concept of a virtual particle may be confusing. The important point is that the electron and photon are represented by quantum fields because of the wave–particle duality. They only have all the properties of real particles when they are in the initial state or detected in the final state. One of the ways of interpreting Figure 4.1b is to say that one electron produces an electric field that the other electron feels. This then reduces to classical electrodynamics giving Coulomb's law. Thus, in this case the virtual photon is equivalent to the electric field. However, as discussed below, it is also necessary to consider the possibility that one electron emits two virtual photons so that the final answer is not the same as classical electrodynamics.

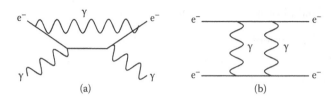

FIGURE 4.4   Feynman diagrams representing higher-order contributions to the processes of Figure 4.3a and b, respectively.

The diagrams in Figure 4.3 are referred to as tree diagrams, which means that they contain the minimum number of vertices needed to explain the different processes. However, there are diagrams for the same processes with more vertices. Some examples are shown in Figure 4.4. Thus, Figure 4.4a produces the same reaction as Figure 4.3a, and Figure 4.4b produces the same reaction as Figure 4.3b.

The key to QED is that diagrams with more vertices, like those in Figure 4.4, give a much smaller contribution to the process than the tree diagrams in Figure 4.3. This is due to the small value of the electron charge, e. Therefore, using only the tree diagrams already gives a good approximation to the correct result, while adding the diagrams with four vertices will make a small improvement, and going to six vertices a very small improvement. This method of calculation is called perturbation theory.*

The truth is not quite this simple. In the 1930s, when physicists tried to calculate higher-order contributions, they found crazy, even infinite, answers. This didn't bother some physicists since the simple tree-level results agreed with available experiments given their limited accuracy. In the 1940s a couple of more precise experimental results were obtained that appeared to deviate slightly from the simple theory. The solution came in the late 1940s with the development of renormalization theory by Feynman, Julian Schwinger, and Sin-itero Tomanaga, who shared the 1965 Nobel Prize. This is discussed in Appendix 2. As a result, it has become possible to carry out perturbation theory with results that agree with the experiment to an amazing level of precision.

4.2 GREGARIOUS PARTICLES AND LONESOME PARTICLES: SPIN AND STATISTICS

Imagine that you have two cars of the same brand, color, etc. They may look very similar, but you can distinguish them by their license plates.

---

* The final result can be expressed as an expansion in powers of $\alpha = e^2/c = 1/137$.

Even if they looked exactly the same, with fake license plates, you could distinguish them because they cannot occupy the same space; one car is over here and the other over there. This is not so with fundamental particles because of quantum mechanics.

Recall that each particle is described by a wave function. Evaluating the square (of the absolute value) of this wave function at some point in space-time represents the probability that the particle will be found at that location at that time. Let's consider two identical particles whose wave functions overlap. We will name them particle 1 and particle 2. In the locations where the wave functions overlap, there will be some nonzero probability of finding particle 1, and also some nonzero probability of finding particle 2. Imagine that a detector is placed in the overlap region and that it registers a hit. Does this hit correspond to particle 1 or to particle 2? We do not know. Because both wave functions have a nonzero value at the detector's location, either particle might have been there. And, because the particles are identical, the detector cannot, even in principle, tell one from the other.

Figure 4.5 shows an analogy. This analogy is far from perfect, but it may help us understand some of the main ideas. The figure shows two boxes: the grey box and the dashed-black box. The boxes overlap in the middle section. These boxes represent overlapping wave functions. Now we place one particle in each box. For example, we place particle 1 in the grey box and particle 2 in the dashed-black box, as in Figure 4.5a. If a detector placed in the overlap region finds a particle, we do not know whether it is particle 1 or particle 2. Worse, we cannot even disentangle the situation

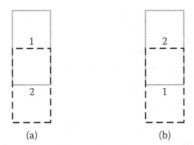

FIGURE 4.5   Picture evoking a situation where particles 1 and 2 occupy two states. One state is represented by the grey box; the other state is represented by the dashed-black box. (a) Particle 1 is in the grey state, and particle 2 is in the dashed-black state. (b) Particle 2 is in the grey state, and particle 1 is in the dashed-black state.

described in Figure 4.5a from that described in Figure 4.5b, in which we have placed particle 1 in the dashed-black box and particle 2 in the grey box. Because we cannot disentangle the two situations, the wave function of the system should be represented by a superposition of both possibilities. How is this superposition to be accomplished?

When the particles are indistinguishable, quantum mechanics allows two possibilities for the wave function of the system. These are represented schematically in Figure 4.6. Let's explain what we mean by Figure 4.6a in detail. We are supposed to do the following. Find the wave function describing particle 1 in the grey box. Find the wave function describing particle 2 in the dashed-black box. Find the product of the two. This is the wave function of Figure 4.5a. Then, find the product of the wave function describing particle 1 in the dashed-black box and the wave function describing particle 2 in the grey box. This is the wave function of Figure 4.5b. Finally, add the wave function of Figure 4.5a to the wave function of figure 4.5b.

We now notice something very important. Because the two particles are indistinguishable, labeling them 1 and 2 has no physical meaning. We could equally well have named them in the reverse order: 2 and 1. If you exchange the labels 1 and 2 in Figure 4.5a, you get Figure 4.5b, and vice versa. As a result, if you exchange 1 and 2 in Figure 4.6a, you get exactly the same. We say that the wave function of the system is symmetric under $1 \leftrightarrow 2$ interchange. Similarly, if you exchange 1 and 2 in Figure 4.6b, you get the same figure, but multiplied by an overall minus sign. We say that the wave function of the system is antisymmetric under the $1 \leftrightarrow 2$ interchange. Quantum mechanics tells us that the wave function of two indistinguishable particles is either symmetric or antisymmetric. There is no other possibility.[2]

There is an exact correlation between the type of wave function of a system of indistinguishable particles and the particle's spin. The wave function of fermions (half-integer spin) is always antisymmetric under particle interchange; the wave function of bosons (integer spin) is always symmetric under particle interchange.

This has several interesting implications. We start with two fermions. Their wave function is pictured in Figure 4.6b. Imagine that we decide to place the two fermions in the same grey box. The result can be found by coloring with solid grey the dashed-black box in Figure 4.6b, that is, by making both boxes grey. But then we are subtracting a grey-grey wave function from itself; the result is zero! This means that it is impossible for two fermions to have exactly the same wave function. Fermions like to

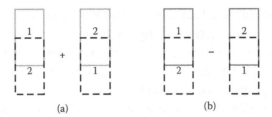

FIGURE 4.6 Picture evoking a situation where particles 1 and 2 occupy two states in (a) a symmetric fashion and (b) an antisymmetric fashion. In Figure 4.6b the overlap region cannot be occupied.

have their own space. Notice that this has nothing to do with the interactions between the fermions, such as electric repulsion. Neutrons and neutrinos have zero charge, and satisfy this condition. This property is a pure quantum effect with no classical analogue. It was first identified by Pauli in 1924 when studying the electrons in atoms. There, for each orbital configuration there are only two possibilities for the spin: spin up or spin down. Thus, we may have at most two electrons in the same orbital. This is known as the Pauli exclusion principle.

Many properties in nature can be explained by stating that all systems would prefer to be in the configuration corresponding to the minimum energy. One example can be found by letting a marble roll inside a bowl. As friction slows the marble, it will come to rest at the lowest possible point within the bowl. Electrons in atoms would also like to occupy the orbital of least energy. But, since the antisymmetric nature of the wave function implies their mutual avoidance, only two electrons can be in this orbital: one with spin up, the other with spin down. Any extra electrons must be placed in higher orbitals. This simple picture allowed Pauli and his contemporaries to understand many features on Mendeleev's periodic table of chemical elements.

If fermions are loners, bosons are gregarious. In fact, if you paint the dashed-black box in Figure 4.6a grey, you get the sum of a grey-grey wave function with itself. No zero, no problem. Thus, there is the possibility that bosons might all try to occupy simultaneously the lowest possible energy state in a system, having zero kinetic energy. This possibility, which was predicted by Einstein in 1924–1925 based on a counting method proposed by Bose, is known as Bose–Einstein condensation. Because temperature is related to particle motion, extremely low temperatures are needed. It was only 70 years after its prediction that Eric Cornell, Wolfgang Ketterle, and Carl Wieman would achieve the 0.00000002 degrees above absolute zero needed for this discovery (slightly below –273°C or –459°F). They were awarded the 2001 Nobel Prize "for the achievement of Bose–Einstein condensation in dilute gases of alkali atoms, and for early fundamental studies of the properties of the condensates."[3]

How is this related with quantum fields? Everyone knows that two times three is the same as three times two (six in both cases). And this is true for any two numbers you might choose; you change the order of the multiplication, but you still get the same result. Scientists refer to this fancifully by stating that the multiplication of two numbers is commutative. Stating that the numbers commute means that multiplying in one order is equal to multiplying in the other. However, not all things commute. Let us think of two operations: closing the drawer and placing the socks inside. Of course, you can place the socks inside and then close the drawer. But you certainly cannot do the opposite. We would say that these two operations do not commute, since one of the combinations is not even possible.

The act of creating two bosons shares with ordinary numbers the commutation property. That is, we can create boson 1 and then create boson 2, and this gives the same result as creating boson 2 and then creating boson 1. This does not happen with fermions. The act of creating fermion 1 and then creating fermion 2 gives a Feynman diagram that has a minus sign with respect to first creating fermion 2 and then creating fermion 1. This is the quantum field theory origin of the wave function's symmetry properties that we found in ordinary quantum mechanics.

NOTES

1. To be precise, an amplitude is a complex number, and the corresponding probability is proportional to the square of the magnitude of the amplitude. Complex numbers are explained in Appendix 3. Amplitudes are also known as probability amplitudes.
2. In our simplified example, we considered only the spatial wave function. In general, the full wave function of a particle also includes pieces describing the internal spaces, such as spin and color (the charge of the strong interaction). The full wave function will be given by the product of the various wave functions. It is the full wave function that must have the symmetry properties that we have described. For example, a system with three fermions can be symmetric in space and symmetric in spin, as long as it is antisymmetric in color. This is actually how the color degree of freedom was first introduced in connection with the baryon Δ^{++}.
3. Source: Nobel Foundation–nobelprize.org.

CHAPTER 5

The Birth of Particle Physics

Pauli's Neutrino, Fermi's Weak Interaction, and Yukawa's Pion

5.1 ELECTRON, PROTON, AND NEUTRON

Since ancient times people have speculated on whether there is a smallest piece of matter. A variety of experiments in the late nineteenth and early twentieth centuries identified the atom as the smallest piece of a chemical element.

The atom turned out to be very small, so that 100 million lined together would take up only 1 or 2 cm. In 1895 Thomson discovered a constituent of atoms, the electron, which had a negative charge and was much lighter than a single atom. Since atoms were electrically neutral, presumably nearly all the mass of the atom was in a positively charged piece.

In 1913, Rutherford performed an experiment in which a small positively charged alpha particle (discussed below) was scattered from a piece of gold. The large repulsive force that was observed indicated that the alpha particle had come very close to an intense positive charge. In this way it was deduced that an atom consists of a very small positively charged nucleus surrounded by electrons. The radius of the nucleus was ten thousand to one hundred thousand times smaller than that of the atom. It was determined that different chemical elements had different numbers of electrons in the atom. This is called the atomic number, Z. The charge of all electrons is $-Ze$, and the charge of the nucleus is $+Ze$, making the atom electrically neutral.* The value of Z varies from 1 to 92, as we go from hydrogen to uranium in the periodic table.

The nucleus of hydrogen, which has an equal and opposite charge to that of the electron, is called the proton. But what about the nuclei of other atoms? For example, the nucleus of helium has Z = 2, but its mass is about four times that of hydrogen. This problem was solved with the discovery of a neutral particle, called the neutron, which has a mass very close to that of the proton. Thus, it was concluded that nuclei consisted of Z protons, and the rest of the mass was made up of neutrons. For example, the nucleus of helium contains two neutrons and two protons.

In fact, a given chemical element with Z protons in the nucleus can occur in different forms, called isotopes, which differ in the number of neutrons. For example, lithium, which has Z = 3, has two isotopes ^{6}Li with three neutrons and ^{7}Li with four neutrons. The number of neutrons plus the number of protons (six or seven in the case of lithium) is called the mass number.

Thus, we are left with a simple picture: all matter is made of three constituents—electron, proton, and neutron. The electrons are bound to the nucleus by the electrical force; in the simple Bohr model they rotate in orbits around the nucleus like planets around the sun. There must be some new strong interaction that holds the neutrons and protons closely together in the nucleus in spite of the repulsion between the protons. We now turn to the discoveries that took us beyond and eventually far beyond this simple picture.

5.2 BETA DECAY AND PAULI'S NEUTRINO

In 1895, Henri Becquerel was studying a phenomenon called photolumi-nescence. He exposed crystals containing uranium to sunlight and then detected rays that were emitted as a result of the absorbed sunlight energy. During a week of cloudy weather he had to leave his unexposed crystal and photographic plate (used for detecting the rays) in his desk drawer. He found to his amazement that the photographic plate was full of tracks. The uranium was spontaneously emitting rays without ever having been exposed to the sun. He had discovered radioactivity by accident.

Soon after, a young Polish student named Marie Slodovska came to Paris and became interested in radioactivity. She obtained a large sample

---

* We use the usual convention that e refers to the charge of the positron. Thus, the charge of the electron is −e.

of uranium ore, called pitchblende, and performed a chemical analysis. She discovered there were elements besides uranium in the ore, in fact, elements that had never been seen before. The element with Z = 84 she named polonium after her native country. It was more radioactive than uranium; that means that it emitted more rays for a given mass. Then she discovered an even more radioactive element with Z = 88, radium. In 1903, Marie Curie (her married name) received the Nobel Prize in Physics. Sad to say, it was 60 years before another woman (Maria Goeppert-Mayer) won the prize.

What were these emitted rays? Ernest Rutherford studied them by bending them in a magnetic field. He found three types, which he labeled alpha, beta, and gamma. The beta rays consisted of electrons, the alpha particles appeared identical to the nuclei of helium, and the gamma rays appeared to be like high-energy x-rays. The alpha particle could be understood as a piece of the nucleus that broke off; the mechanism for this (called tunneling) was only understood after the development of quantum mechanics. The gamma rays could be understood as photons emitted when a nucleus went from an excited state to the ground state, analogous to the emission of light by an atom. The origin of the beta rays was more mysterious. Rutherford discovered that the electrons appeared to come from the nucleus where a neutron had changed into a proton. Several radioactive decays led to a sequence of different nuclei ending with an isotope of lead, as shown in Table 5.1. The alphas and the gammas had discrete energies

TABLE 5.1   Decay Chain, Ending with Stable Lead

Nucleus	$^{238}U_{92}$	$^{234}Th_{90}$	$^{234}Pa_{91}$	$^{234}U_{92}$	$^{230}Th_{90}$
Half-life	5×10^9 years	24.1 days	1.18 min	2.5×10^5 years	8×10^4 years
Ray emitted in decay	α	β, γ	β, γ	α	α, γ
Nucleus	$^{226}Ra_{88}$	$^{222}Rn_{86}$	$^{218}Po_{84}$	$^{214}Pb_{82}$	$^{214}Bi_{83}$
Half-life	1,620 years	3.82 days	3.05 min	26.8 min	19.7 min
Ray emitted in decay	α, γ	α	α	β, γ	β, γ
Nucleus	$^{214}Po_{84}$	$^{210}Pb_{82}$	$^{210}Bi_{83}$	$^{210}Po_{84}$	$^{206}Pb_{82}$
Half-life	1.7×10^{-4} s	19.4 years	5 days	1.38 days	Stable
Ray emitted in decay	α	β, γ	β	α, γ	

Note: The subscript number to the right of the nucleus' name gives the number Z of protons inside the nucleus. The superscript number to the left gives the mass number (number of protons plus neutrons). The half-life is the time necessary for the decay of half of the initial nuclei.

corresponding to the energy difference between the initial and final states. However, after many experiments it was determined that the electrons had a continuous distribution of energies. This meant that some electrons could emerge with any energy above zero and below some maximum value. How could this happen when there was a transition from one nuclear state of definite energy to another? In desperation, Niels Bohr even suggested that at the nuclear level energy conservation was violated. In 1930 the famous physicist Wolfgang Pauli wrote a letter to a convention of physicists suggesting a solution. He proposed that along with the electron there emerged a second electrically neutral particle that no one had ever detected. When an electron emerged with less than the maximum energy, the rest of the energy was carried by this new particle. He admitted that this was a rather wild idea, but that "only those who wager can win." In fact, Pauli did not bet very much on this idea, since he only wrote about it in this letter and did not publish an article about it. He gave this particle the name *neutron*. After the particle we all know as the neutron was discovered, someone asked Enrico Fermi what was the difference between Pauli's neutron and the other neutron, and Fermi said that Pauli's was "a leetle neutron, a neutrino," and so it has been ever since.

5.3 FERMI'S WEAK INTERACTION

The question arose: What was the origin of the electron and neutrino? The electron somehow came from the nucleus since the nucleus changed from one element to another. The nucleus could be explained as consisting of neutrons and protons; it made no sense to have electrons inside it. The answer given by Fermi was that the electron and neutrino were created from the available energy exactly as a photon was created when an atom emitted light. In the case of the photon the emission is due to the electromagnetic interaction, but what was responsible for the emission of the electron and neutrino? The answer given by Fermi was that there was a new interaction.

In 1933 Fermi invented the weak interaction, a new fundamental interaction. In Figure 5.1a we show Fermi's interaction diagrammatically; it is to be compared with the electromagnetic interaction shown in Figure 4.1a. In Figure 4.1a, at a point in space an electron makes a transition from one state to another, and a photon emerges and there is a coupling strength e. In Figure 5.1a, at a point in space a neutron in one state makes a transition to a proton in another state, and an electron and an antineutrino emerge. (Notice that the convention now is to call the particle emitted together with the electron an antineutrino in analogy with the electrodynamic process in which an electron is emitted together with an antielectron, the positron.) The associated coupling strength is given by the Fermi constant, G_F, which Fermi determined from the rate of beta decay. The theory that describes the emission of the photon also describes other processes, as shown in Figures 4.1 and 4.2. Figure 4.1b is the time reversal in which a photon is absorbed. Similarly, Fermi's theory allows for the prediction of other processes, in particular, the absorption of an antineutrino represented in Figure 5.1b, that is, the rate at which antineutrinos interact with matter.

Knowing the value of G_F, it was possible to calculate the rate, the cross section, for this interaction. The answer was exceedingly small. If a million antineutrinos from beta decay were to enter the earth, all except perhaps one would emerge on the other side without interacting. It seemed clear that Pauli's particle would never be detected. Figure 5.2 shows analogous processes for neutrinos.

The decay process shown in Figure 5.2a was not discovered using natural radioactive elements like those in Table 5.1. However, new isotopes that

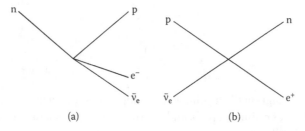

FIGURE 5.1   (a) Diagram for the beta decay process: neutron goes to proton plus electron plus antineutrino. (b) Diagram for the neutrino interaction: antineutrino plus proton go into positron plus neutron.

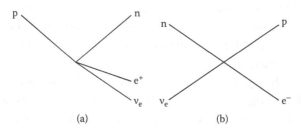

FIGURE 5.2   Diagrams analogous to those in Figure 5.1, but involving neutrinos rather than antineutrinos: (a) proton goes to neutron, neutrino, and positron; (b) neutrino plus neutron goes to proton plus electron.

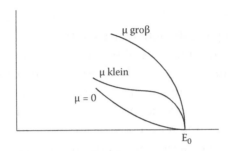

FIGURE 5.3 Fermi's prediction for the spectrum dependence on the neutrino mass. (Taken from Fermini's "Versucheiner theorie der postrahlen—An attempt of a theory of beta radiation." *Z. Phys.* 88 (1934), fig. 1, 171. With permission.)

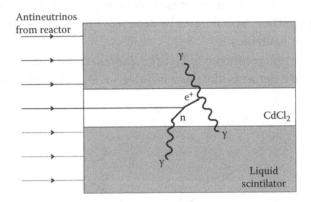

FIGURE 5.4 Diagram illustrating Reines' apparatus. The neutron, n, is captured by the nucleus of cadmium (Cd), and the three final gamma rays are detected by the liquid scintillator.

decay with the emission of positrons were produced in the laboratory as the result of nuclear collisions. This was referred to as artificial radioactivity, discovered in experiments by Irène Joliot-Curie and her husband, Frédéric Joliot. For this they were awarded the Nobel Prize in Chemistry in 1935; unfortunately, her mother, Madame Curie, had died 2 months earlier. Isotopes that emit positrons have a variety of practical uses, particularly in medicine.

Using his theory, Fermi calculated the continuous spectrum of electrons and found that it agreed with the experimental results provided the "rest mass of the neutrino is either zero or at any rate much smaller than the mass of the electron." Figure 5.3, taken from Fermi's original article, shows his prediction for the spectrum in case the neutrino had a mass. It is only in recent years that we finally have evidence for a nonzero very small mass for the neutrino. In 1955, 25 years after Pauli's letter and 2 years before Pauli died, Fred Reines and Clyde Cowan detected the antineutrino, as illustrated in Figure 5.4. A number of technological developments made this possible, especially the discovery of nuclear fission, which led to the construction of nuclear reactors. A reactor produces large numbers of neutron-rich isotopes that are extremely radioactive; thus, there is a large flux of antineutrinos. As shown above, the antineutrino interaction produces a positron and a neutron. In the experiment the positron annihilates with an electron to yield two gamma rays, while the neutron is captured by a nucleus that goes to an excited state and then emits a gamma ray. The detection of these three gamma rays uniquely identifies the antineutrino. It was only 40 years later that Fred Reines won the Nobel Prize for the detection of the neutrino.

The neutrino that once seemed mythical has become increasingly important as a probe of fundamental physics, and neutrinos from the sun and from an exploding star have provided unique insights in astronomy. The amazing story of the neutrino is the subject of Part C. The neutrino was the first fundamental particle that had nothing to do with the constituents of matter, but many more were to follow.

5.4 NUCLEAR FORCES AND YUKAWA'S PION

In 1935, Hideki Yukawa proposed a theory of the strong interaction that binds protons and neutrons together in the nucleus of the atom. Like Fermi, he was looking for an analogy with the electromagnetic interaction. For the interaction between an electron and a proton, one may consider the proton as producing an electric field that reaches the electron and causes the attraction. This can be represented by analogy with Figure 4.3b of Chapter 4 as involving a virtual photon. Yukawa proposed that a proton could turn into a neutron producing a field that had a positive charge. When this field reached a neutron, it changed it to a proton and caused an attraction between the neutron and the proton. Associated with this field there would be a charged particle, which has come to be called the pion. This can be shown as in Figure 5.5a, where the dashed line is a virtual pion.

The problem was that if you assumed this pion had zero mass like the photon, then the force would vary with distance as $1/r^2$. However, the force between the neutron and proton has to have a very short range; that is, it must fall very rapidly with distance. This is necessary in order to explain why the neutron and proton are held so closely together. Yukawa's

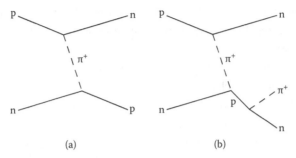

FIGURE 5.5 (a) Proton-neutron interaction via a charged pion. (b) Pion production in proton-neutron interaction.

discovery was that this required the particle associated with the field to have a mass; the larger the mass, the shorter the range. In order to explain the observed size of nuclei, the mass had to be about two hundred times that of the electron, which is about one-tenth that of the proton.*

Given the vertex, or the coupling, that connects the neutron, proton, and pion (as seen in Figure 5.5), it should be possible to produce a pion if a proton or neutron is given enough energy, just as photons are emitted by electrons. Figure 5.5b is a diagram illustrating the production of a pion resulting from a collision between a proton and a neutron. Such a collision has to be quite energetic in order to provide the energy Mc^2, where M is the mass of the pion. Before the development of synchrocyclotrons in the 1950s, the only place to observe such collisions was in the cosmic rays, which consist of high-energy protons that enter the earth's atmosphere from outer space. While such collisions were not directly observed, physicists studying cosmic rays discovered charged particles with a mass intermediate between that of the electron and the proton. So it seemed Yukawa's pion had been discovered in 1937.

5.5 THE MUON: WHO ORDERED THAT?

Further observations showed that this particle decayed, yielding an electron. However, a problem arose. This particle did not seem to have strong interactions with nuclei. If this particle is produced strongly, it should also be absorbed strongly. The answer came with pictures taken using photographic emulsions atop the Pyrenees. The type of event observed is illustrated in Figure 5.6. The incoming track is actually the pion, which then stops and decays into a new charged particle, and that particle stops and

---

* The relation between the range and the pion mass is discussed in Appendix 5.

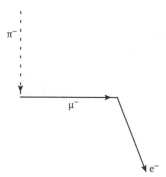

FIGURE 5.6 Type of event observed studying cosmic rays in the Pyrenees in 1947.

decays into an electron. It was this new particle that had been seen before. Because the pions decay soon after they are produced in the atmosphere, the particles that were observed at sea level were these new particles, which live much longer than the pion.

What are these new particles? Yukawa said there should be a pion, but as the famous physicist I. I. Rabi asked, "Who ordered this new particle?" It was a charged particle and, like the electron, it had no strong interaction, but it had a mass two hundred times that of an electron. It could be considered a heavy brother of the electron. It came to be called the muon and is usually represented by the Greek letter mu (μ).

Looking at Figure 5.6, we see a muon emerging from a pion at rest, but nothing emerging on the other side to conserve momentum. It didn't take a Pauli to shout "neutrino"; the pion decayed to a muon and a neutrino.

5.6 THE MUON NEUTRINO: A NEW KIND OF NOTHING

After the experiment of Reines, the question arose whether this neutrino could be detected. With the development of accelerators that produced high-energy beams of protons, it was possible to produce lots of pions by bombarding nuclei with these protons. It was then pointed out by Mel Schwartz that, as these pions decayed, you could have a flux of neutrinos and study their interactions, just as Reines had done. In 1962, the experiment was carried out at the Cosmotron accelerator at Brookhaven National Laboratory. Interactions were observed, but there were no electrons coming out as in the Reines experiment; rather, the interactions produced muons. This was a new kind of neutrino, presumably a brother of the other neutrino, just as the muon was the brother of the electron. The *New York Times* headline read: "Physicists Discover a New Kind of Nothing."

We now distinguish the two neutrinos as v_e (pronounced "nu-e," the electron neutrino) and v_μ (pronounced "nu-mu," the muon neutrino). For the discovery of this second neutrino, Schwartz, Jack Steinberger, and Leon Lederman won the Nobel Prize in 1988. Why they won the prize for the second neutrino before Reines won his for the first neutrino, we cannot explain.

5.7 STRANGE PARTICLES

In 1950 there was announced the discovery of V particles in cloud chambers observing cosmic rays. What was observed were two tracks that appeared like an upside-down V. One track was a proton and the other a negative pion. Later this came to be called the lambda particle, since the capital Greek letter lambda (Λ) is an upside-down V. This is a neutral particle that decays into a proton and a negatively charged pion. If this decay were the result of a strong interaction, the decay lifetime would be extremely short, so they would decay very soon after they were produced. But, in fact, it appeared that they had traveled an average distance around 3 cm in the cloud chamber before decaying. This led to a strange problem: If these particles lacked a strong interaction, how come they seemed to be abundantly produced? The answer proposed by Murray Gell-Mann was that these particles had a new type of charge, denoted by S, that was conserved in the strong interaction but violated in the weak interaction. In particular, lambda had S = –1. In this case, the lambda could be produced in the strong interaction in association with another particle that had S = +1. Finally, in 1954 it was possible to study the production process at the Brookhaven Cosmotron by sending beams of pions into a sensitive type of cloud chamber. It was seen that together with the lambda, a second particle was produced that came to be called the kaon and that decayed into two pions. Presumably the kaon had S = +1. This new type of charge, S, came to be called strangeness, and the lambda and the kaon were strange particles.

The different particles could be divided into two types. There were the leptons, which had no strong interactions; these were the electron and muon and the two neutrinos. The others that had strong interactions were called hadrons; these include the proton, neutron, lambda, pion, and kaon. As the 1950s went on, more new hadrons were discovered. Charts of particles were produced that began to look like the periodic table of the elements. Was there some substructure underlying this "particle zoo"?

CHAPTER **6**

Learning to Live with Gell-Mann's Quarks

6.1 ORIGIN OF THE QUARK THEORY

In the early 1950s, experiments were carried out using synchrocyclotrons in which beams of pions were scattered from protons. It was discovered that the scattering became large at a certain energy; this is referred to as a resonance. In the scattering of a photon from an atom, a resonance occurs if the energy is just right to produce an excited state of the atom. Essentially, the photon is absorbed to yield the excited state, and then the excited state quickly emits back the photon, leaving behind the initial atom in its unexcited state (ground state). Thus, it appeared that in the case of the pion-proton scattering, an excited state of the proton had been discovered.

Remembering the equivalence between mass and energy, we can describe the excited state as having a mass greater than that of the proton. This state was given the name delta (represented by the Greek letter Δ^0), and the mass was about 1.3 times the proton mass. For the first experiments using negative pions the delta was neutral; thus, you might call it an excited state of the neutron. A resonance with practically the same mass was found using positive pions, thus producing a doubly charged delta (Δ^{++}). Further experiments with scattering from neutrons discovered two other charged states: delta plus (Δ^+) and delta minus (Δ^-).

As still more resonances were discovered and more strange particles identified, the idea that there were some more fundamental particles that were the constituents of the proton and the neutron seemed very interesting. In 1964 two papers appeared with a proposal for these. One was a short paper by the famous physicist Murray Gell-Mann; the other was a detailed report by a young physicist named George Zweig. Gell-Mann called the new particles quarks, after an obscure passage in James Joyce's *Finnegans Wake*; Zweig simply called them aces. Gell-Mann's name won out; Zweig got a job at the California Institute of Technology.

The neutron and proton consisted of two types of quarks: u (the up quark) and d (the down quark). The u quark had 2/3 of the charge of the proton, and the d quark a charge of –1/3 that of the proton. The proton consisted of three quarks (u, u, and d), and the neutron of a different combination (d, d, and u). The quarks had spin 1/2 like the neutron and proton, so that in the proton the d spin was opposite in direction to that of the spins of the u's. The neutron is obtained by interchanging d and u. Now it is easy to explain the four charge states of the delta, as shown in Table 6.1. Furthermore, it was discovered that the delta had spin 3/2, so that all the quark spins were aligned.* Presumably, the difference in mass was related to the interaction energy between the spins.

The pions could now be understood as a quark combined with an antiquark. Antiquarks are represented by placing a bar over the letter representing the corresponding quark. For example, the d antiquark is represented by $\bar{d}$, and is sometimes referred to as dbar. Its charge is +1/3, opposite that of the d quark. Now, consider the combination $u\bar{d}$. Its charge

TABLE 6.1   Quark Content and Approximate Mass of Some Hadrons

Symbol	Particle Name	Quark Content	Mass (MeV)
π^+	Pion	$u\bar{d}$	139.6
π^0	Pion	$d\bar{d} - u\bar{u}$	135.0
π^-	Pion	$\bar{u}d$	139.6
p	Proton	uud	938.3
n	Neutron	udd	939.6
Δ^{++}	Delta	uuu	1,232
Δ^+	Delta	uud	1,232
Δ^0	Delta	udd	1,232
Δ^-	Delta	ddd	1,232

---

* Three u quarks with the same spin direction would seem to violate Pauli's exclusion principle. As explained in Section 6.4, this problem was solved by quantum chromodynamics (QCD) by saying that the quarks had different color charges.

is given by the sum of the charge of the up quark (+2/3) with the charge of the down antiquark (+1/3); we obtain the same charge as the proton, and this is the positively charged pion, denoted by π^+. Its antiparticle is π^-, with quark content $\bar{u}\,d$ and charge –1.

As shown in Table 6.1, it is natural, in addition to the charged pions, to have a neutral pion. This was discovered in 1950; it decays into two photons as the quark and antiquark annihilate. Actually, there are two neutral possibilities: u quark u antiquark and d quark d antiquark; the neutral pion (π^0) is understood to be a combination of the two. Another neutral meson was found, the eta (represented by the Greek letter η), with a larger mass that is understood to be the other combination.

The mass units used in Table 6.1 are common among particle physicists, but they correspond to an abusive use of the energy units eV (electron-volt) to describe mass. The abuse arises from Einstein's $E = mc^2$ formula, where c is the velocity of light. Strictly speaking, one should express the mass of the proton as 938.3 MeV/c². In turn, the eV refers to the energy gained by an electron accelerated by a 1 V battery, and the prefix M stands for million. So, the proton mass is 938,300,000 eV/c², which is approximately 1.67×10^{-27} kg. A very small mass indeed! In the usual notation, the c² is dropped and we express the masses of fundamental particles in keV = 1,000 eV; MeV = 1,000,000 eV; or GeV = 1,000,000,000 eV.

Following the description of protons and neutrons in terms of quarks, the weak interaction could then be expressed as a vertex connecting the d and u quarks to the electron and the electron antineutrino. Figure 6.1 shows that the neutron changes to a proton because a d quark changes into a u quark. Similarly, there could be a vertex connecting u and d to $\mu^-\bar{\nu}_\mu$. This could explain the decay of the pion. Instead of a d going to a u, we have a d quark and a u antiquark coming together to produce $e^-\bar{\nu}_e$ or $\mu^-\bar{\nu}_\mu$. This is

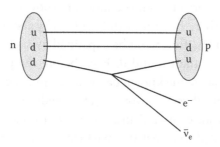

FIGURE 6.1   Quark transition diagram driving neutron decay.

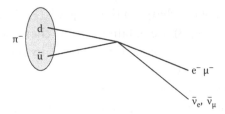

FIGURE 6.2   Quark transition diagram driving charged pion decay.

shown in Figure 6.2. In fact, the decay into an electron is highly suppressed relative to that into a muon, for reasons to be explained later.

To explain the strange particles, a third strange quark, s, was introduced with S = –1 and charge –1/3. Thus, the lambda particle consisted of sud. While u and d have only a small mass difference, the s quark is heavier, explaining why the lambda is heavier than the neutron and proton. The combination of three quarks suu and sdd forms the sigma+ (Σ^+) and sigma– (Σ^-), which are slightly heavier than the lambda. There is also another way of combining sud to form the neutral sigma (Σ^0). The kaons with S = –1 are the combinations of an s quark with a u antiquark and of an s quark with a d antiquark. The Λ and Σ strange particles decay weakly to neutrons or protons (plus a pion). To explain this, there must be a weak interaction that changes an s quark to a u quark; this is discussed in Chapter 9.

As more experiments were done, particles were found that appeared to have S = –2. These were named xi (represented by the Greek letter Ξ). Thus, there was the Ξ^-, which decayed weakly into $\Lambda + \pi^-$ with $\Delta S = 1$ (that means, changing S up by one unit), analogous to the decay $\Lambda \rightarrow p + \pi^-$. In the quark model, Ξ^- would be made up of s and d quarks. There was also a Ξ^0 that decayed into $\Lambda + \pi^0$. Particles made of three quarks have strong interactions and are collectively known as baryons; examples of baryons with different numbers of strange quarks are shown in Figure 6.3. The Σ and Λ are heavier than the proton and neutron by about 20%, and the Ξ is heavier by about 40%. This could be explained by saying that the s quark is somewhat heavier than u and d. Even before the proposal of quarks, attempts to systematize the known particles led to the prediction of a state with S = –3. In 1963 physicists working at Brookhaven discovered a particle that decayed weakly into a Ξ and a pion. On the basis of its production and decay, it was deduced that this particle had S = –3. This was clearly the state sss made up of three strange quarks, as would be expected in the

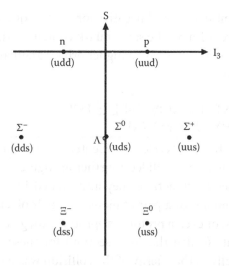

FIGURE 6.3 Quark content of spin 1/2 baryons. The vertical axis is the strangeness S. The baryons shown are the lightest ones for a given value of S. The horizontal axis, I_3, is proportional to the number of u quarks minus the number of d quarks.

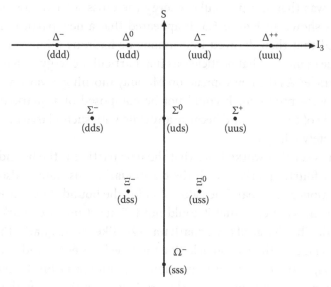

FIGURE 6.4 Quark content of spin 3/2 baryons.

quark model. Its mass was higher than that of the Ξ, which had only had two s quarks, and it decayed by the change of one s quark to a u quark. It was named omega minus (represented by Ω^-).

Figure 6.4 shows the quark content for the ten lightest baryons of spin 3/2. In these figures, S refers to the strangeness content. Conventionally, it is positive for s antiquarks and negative for strange quarks s. The Σ and Ξ baryons in Figure 6.4 have the same quark content as those in Figure 6.3 (hence the same name), but have a higher orbital angular momentum and thus a higher mass.*

6.2 A WEEKEND IN NOVEMBER: THE DISCOVERY OF CHARM

At the Stanford Linear Accelerator Center (SLAC) in California, electrons and positrons were collided together at high energy. As a result of the electromagnetic interaction, they annihilated into a variety of final states: a pair of muons, or a pair of pions, or a pair of kaons, or a number of pions or kaons, or even a proton-antiproton pair, given the large energy available. The rate (called the cross section) for these various processes increased gradually as the energy of the collision was increased.

One weekend in November 1974 experimenters decided to vary the energy in very small steps centered around a particular energy where they had previously seen an indication of something unusual. What they discovered was that at a particular energy the cross section rose to a large value, as shown in Figure 6.5. It appeared that a new particle had been produced that then decayed into muons or hadrons (pions and kaons). The value of mc^2 for that particle was that particular energy. What was this new particle? As the news spread on Monday morning, a variety of explanations were proposed. It could not be composed of a quark-antiquark pair made of the u or s or d because any state with such a large mass would immediately fall apart.

The answer that worked was that the new particle is the bound state of a heavier fourth quark, c, with the c antiquark. This state is also known to practitioners as ccbar. The value of m for the bound state was less than twice the mass of c, so that it could not fall apart into a c quark and a c antiquark. The charge of the c quark was +2/3 like the up quark. The bound state was labeled the psi particle (for the Greek letter Ψ) and decayed by electromagnetic interaction into muons or, more probably, by the strong interaction into pions or kaons. The production of the psi by the electromagnetic interaction is illustrated by the diagram in Figure 6.6. Because the c quarks and the c antiquark are coupled to the virtual photon, the psi has to have spin 1.

---

* In the literature, the different sigmas and xis are differentiated by labeling them with their mass values. Thus, the spin 3/2 sigma is labeled sigma (1385).

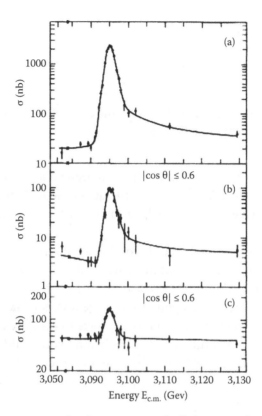

FIGURE 6.5   Cross-section for the transition of colliding e^+ and e^- as a function of the total energy between 3.05 and 3.13 GeV. The mass of the psi is given by the peak at 3.095 GeV. The width of the peak is related to the lifetime of the psi, as discussed in Appendix 5. (J.E. Augustin et al., Quantum Numbers and Decay Widths of the Psi(309s) *Physical Review Letters 34*, 1357 (1974), 1357, fig.1. With permisssion.)

At the same time, evidence for this particle was presented based on its production using a proton beam on a nuclear target at the Brookhaven National Laboratory (BNL). For the discovery of this new particle, Burt Richter of SLAC and Sam Ting working at BNL shared the 1976 Nobel Prize. Since each group gave the particle a different name, the particle is sometimes known J/Ψ (pronounced jay-psi).

Given the large mass of the c quark, it became possible to try to use nonrelativistic quantum mechanics to calculate the bound states of the ccbar system, assuming some kind of binding force. There could be an excited state of spin 1, and such a state was then discovered in the electron-positron collisions. This state, labeled the psi prime, decayed back to the psi by emitting two pions. Gradually a whole set of states was predicted

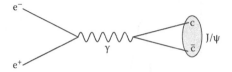

FIGURE 6.6 Feynman diagram describing the $e^+ e^-$ collision, which produces a bound of a c quark with a c antiquark.

and found. This resembled closely what had been done for atoms and molecules 40 years earlier. The success of the quark model in making predictions about charmed particles helped to convince everyone that quarks were here to stay.

There were also bound states involving a c quark and a u antiquark or a c quark and a d antiquark; these were labeled D^0 and D^+ and their antiparticles $\bar{D}^0$ and D^-. It was clear that the c quark had a kind of charge called charm, analogous to strangeness, so that in strong or electromagnetic interactions you would have to produce a c quark and a c antiquark. As a result, in e^+-e^- collisions you could produce the pair D^+ and D^- or D^0 and $\bar{D}^0$. This was observed at Stanford as the energy was raised above the energy of $2m_c c^2$, where m_c is the mass of the charm quark.

6.3 ANOTHER VERSION OF THE RUTHERFORD SCATTERING EXPERIMENT

Over many years, Robert Hofstadter at Stanford studied the scattering of electrons from protons. The first experiments at relatively low energies involved purely elastic scattering; that is, the final state was simply equal to the initial state: a proton and electron. The results deviated from what would be expected if the proton were a point charge and showed that the proton had a small nonzero radius. At much higher energies, there was inelastic scattering in which the final state contained a number of pions besides a proton or neutron. In trying to understand why some electrons were scattered at relatively large angles, it was realized by Bjorken and by Feynman that these events occurred as if the electrons had been scattered by point charges inside the proton. Indeed, the data could be fit if those electrons had been scattered from u and d quarks. The electron scattering had provided a way of looking inside the proton, just as the alpha scattering experiment of Rutherford had shown a way of looking inside the atom.

6.4 QUANTUM CHROMODYNAMICS

As discussed above, physicists found the quark theory extremely useful in analyzing a variety of phenomena. However, an important problem remained: no one had ever detected a free quark. There were a number of searches for a particle with a charge of 1/3 or 2/3 that of the proton and electron, but none were ever found. For this reason, even as people used the quark model, they questioned whether there were really quarks. In a lecture in 1972, Gell-Mann himself expressed his doubts, suggesting that quarks might be "fictitious objects in our models."[1]

A fundamental question was the nature of the interaction that held the quarks together in particles like the proton and neutron. If quarks were the fundamental strongly interacting particles, then the basic strong interaction was that between the quarks. The theory, developed by a number of people, was based on an analogy with quantum electrodynamics (QED). The quarks carried some type of charge that produced a field, and associated with that field was a particle analogous to the photon. This particle was labeled the gluon because it served to glue the quarks together.

There were crucial differences compared to QED. There were three types of charges. An artistic physicist named them after the primary colors: red (R), yellow (Y), and green (G).* The charge was referred to as the color and the theory was called quantum chromodynamics (QCD). Unlike the photon, which is electrically neutral, the gluon also had a color charge. On the other hand, the proton and neutron had no color charge, as the colors R, Y, and G of the three quarks were organized into a neutral (color white) combination. Similarly, the pion had no color, as a quark with color

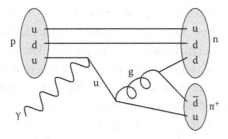

FIGURE 6.7   Pion photoproduction diagram at the quark level. The letters γ, g, p, and n stand for the incoming photon, gluon, proton, and neutron, respectively. Notice the different lines used for photons and for gluons.

---

* The primary colors are really red, yellow, and blue, but the physicist who named them got it wrong. However, as particle physics is concerned these are just names.

R combines with an antiquark with color anti-R. The most important difference between QCD and QED was that the force between two quarks (or between a quark and an antiquark) got larger and larger as the distance between them increased. Thus, you could never have enough energy to remove the quark in a proton away from the other quarks. As shown in the diagram of Figure 6.7, if a high-energy photon were absorbed by a quark in the proton, that energy would end up producing a new quark-antiquark pair; the antiquark would join with the outgoing quark to form a pion, and the new quark would fall back to form a proton or neutron again. Thus, the result would be the photoproduction of a pion. QCD explains why we never found a free quark. The quarks are confined inside particles that have no net color. For proving this confinement of quarks in QCD, David Gross, David Politzer, and Frank Wilczek won the Nobel Prize in Physics in 2004. Since gluons also have a color charge, one can never have a free gluon. The only free particles are those that carry no net color charge.

6.5 JETS

One of the problems with QCD is the difficulty of carrying out quantitative calculations. It is not actually possible to accurately calculate the mass of the proton starting with three quarks or the pion starting with a quark and an antiquark. Because the interaction is so strong, there may be virtual gluons inside the particles. In the case of QED in considering the hydrogen atom, there are effects of virtual photons but these are small corrections that were calculated and observed in the 1940s. In contrast, the corresponding effects in QCD are very large and can only be calculated approximately using large computers.

As the energy increases, a prediction of QCD is that the interaction strength decreases (see Appendix 2). Thus, for interactions at high energy, it is possible to use perturbation theory and calculate using Feynman diagrams. As illustrated in Figure 6.7, the energy given to the quark ends up in an outgoing pion. If the energy is very high, there will be a number of pions. The energy and momentum given to the quark are divided or fragmented into those of the pions. While these pions move in different directions, they all move in the general direction of the original quark. The result is referred to as a quark jet, and the observation of such jets helps to visualize the interaction at the quark level.

In electron-positron collisions at high energy, two jets back to back were clearly observed, as shown in Figure 6.8. The fundamental interaction is understood to be the positron interacting with the electron via the

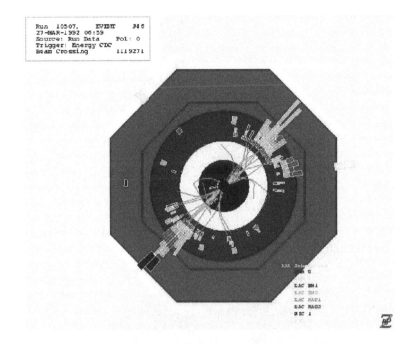

FIGURE 6.8 Picture displaying an event detected at SLAC where an electron and positron pair annihilate, producing two jets. (Courtesy of SLAC National Accelerator Laboratory. With Permission.)

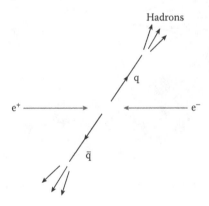

FIGURE 6.9 Schematic representation of an electron-positron collision yielding a quark q and an antiquark.

electromagnetic interaction to produce a quark-antiquark pair, as illustrated by Figure 6.9. The cross section and angular distribution of the two jets agreed with the calculation based on Figure 6.9. As in the case shown in Figure 6.7, the outgoing quarks emit gluons and the gluons produce quark-antiquark pairs. The only particles that can come out freely

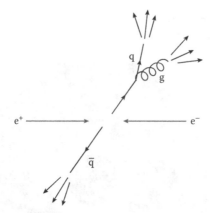

FIGURE 6.10   Schematic representation of an electron-positron collision yielding a quark q and an antiquark and a gluon.

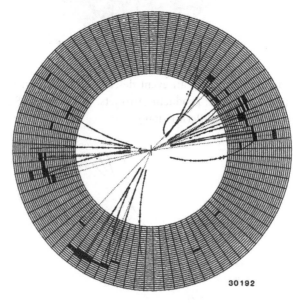

FIGURE 6.11   Figure from DESY 3-jet event. (Source: DESY. With premission.)

are those with no color, like pions, which are the constituents of the jets observed. As a result of the high energy, there are many pions in the jet in contrast to the single pion shown in Figure 6.7.

At higher energies, there is a significant probability that one of the quarks will emit a high-energy gluon, as shown in Figure 6.10. This gluon also must fragment, producing a gluon jet. The result is an event with three outgoing jets. Such events were first observed at the electron-positron collider at the DESY laboratory in Hamburg, Germany, as shown in Figure 6.11. The observed angular distribution of the event agreed with the calculation from QCD. This was considered a major triumph of QCD. As more and more experiments were performed, it became clear that quarks were necessary to understand a whole variety of hadronic phenomena, even though no one would ever detect a free quark.

NOTE

1. Gell-Mann, *Quarks Lecture at Eleventh Meeting on Nuclear Physics at Schladming*, February 21 to March 4 (CERN preprint TH1543, 1972).

CHAPTER 7

Beautiful Symmetries Found and Lost

S INCE ANCIENT TIMES, PEOPLE have found patterns and symmetries in nature. An early example had to do with the stars. Those of us who have lived always amid the light of the city cannot appreciate the grandeur of the sky filled with stars. In ancient times, when the last fire went out, one could not help but gaze at the stars traveling across the sky.

The ancient Greeks proposed that the earth was surrounded by a crystal sphere. The stars were imbedded in this sphere and the whole sphere rotated around once a day. The path of each star formed a perfect circle. This wonderful symmetry was violated by five stars that moved among the others, even sometimes going in the opposite direction from the rest. These were called the planets from the Greek word for "wanderers." They were named after the gods: Mercury, Venus, Mars, Jupiter, and Saturn.

In the sixteenth century, Nicolas Copernicus tried to improve the symmetry. He proposed that the sun and not the earth was the center, an idea that some had already suggested in ancient Greece. The earth was just a sixth planet rotating around the sun, like the other five. The occasional backward motion of a planet simply meant that the earth had passed the planet so that the planet appeared to be going in the opposite direction.

While the Copernicus picture of six planets in circular orbits about the sun agreed qualitatively with observations, it failed in a precise comparison. In the early 1600s Johannes Kepler set out to determine the actual paths of the planets around the sun that would fit the most precise observations of the time, those by Tycho Brahe. He found that the paths were ellipses, not circles. Isaac Newton then showed that the elliptical path followed from his universal law of gravitation. The details of the ellipse depend on the initial conditions, which we now know were set billions of years ago.

Newton's law depends only on the distance between the sun and the planet; it is independent of direction. We say the law of interaction has rotational symmetry. The perfect symmetry is found not in the paths of the planets but in the fundamental laws.

Another interesting symmetry is left-right symmetry or mirror symmetry. This was studied for the case of crystals by the young Louis Pasteur in the nineteenth century. There were crystals of tartaric acid that caused the polarization of a polarized beam of light to rotate in a clockwise direction. There was something in the structure of the crystal that defined a handedness. On the other hand, there were samples of crystals that appeared in their behavior to be identical but which didn't change the polarization. The answer found by Pasteur was that this second sample was really a mixture of two types: one (the same as the first sample) rotated the polarization clockwise, and the other rotated it counterclockwise. The two types of crystals were mirror images of each other.

This mirror symmetry is known technically as parity, and it is a symmetry of the electromagnetic interaction that governs molecules and crystals. Thus, it is possible to arrange the atoms in a molecule so that they have a handedness, but then there must be possible a second molecule with the opposite handedness. The two molecules will have the same binding energy, the same excited states, and appear identical.

When Fermi proposed a theory for the weak interaction, he modeled it after the electromagnetic interaction. As a result, it had the parity symmetry. Then quite surprisingly it was discovered in 1957 that the parity symmetry is violated by the weak interaction.

The most famous example of handedness concerns the chemical constituents of all living things. The basic unit is DNA, which has a left-handed structure. It is possible in the laboratory to synthesize the mirror image DNA. It is possible to imagine living beings made of this right-handed DNA, but no examples have been found on earth. The most plausible explanation is that when these molecules were first synthesized on earth by chance they were left-handed, and that this happened only once more than a billion years ago. There have been attempts to explain the handedness of living things as a result of the parity violation of the weak interaction, but there are no convincing scenarios.

Perfect symmetries and broken symmetries play a very important role in the formulation and application of the theories governing the fundamental interactions among the subatomic particles. As discussed above, these symmetries may not be directly apparent in the world around us, but they underlie it. In the following sections we explain how the physicist understands and applies the concept of symmetry.

7.1 DISCRETE AND CONTINUOUS SYMMETRIES

A symmetry is defined by an operation that leaves something unchanged. We say that something is *invariant* with respect to the symmetry. For example, a rectangle is invariant with respect to a rotation by 180° about an axis perpendicular to the plane of the rectangle. An equilateral triangle is invariant with respect to a rotation of 120° or 240° about the perpendicular axis. The set of operations is known mathematically as a group, and the number of independent operations is the number of elements of the group. The case of the 180° rotation has two elements: the identity (doing nothing) and the 180° rotation. Applying the 180° rotation twice yields the identity. Similarly, the case of a 120° rotation has three elements. Carrying out two of these operations in succession is called multiplying two of the elements together, and the result is another element of the group.

We can extend this to three dimensions. Thus, if we have a box, we can define axes perpendicular to each face of the box and consider rotations about each of these axes, as shown in Figure 7.1. Thus, in the case of a cube there would be symmetry with respect to a 90° rotation about each of these axes. There is an interesting difference in this case. This can be illustrated by applying two rotations about different axes in succession corresponding to multiplying two elements of the group, as illustrated in Figure 7.2.

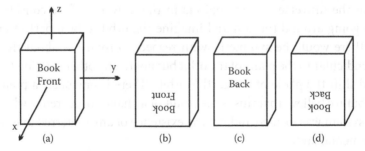

FIGURE 7.1   (a) Initial position of the book. The other figures represent the position of the book after we apply to the initial position a 180° rotation about: (b) the x axis, (c) the z axis, and (d) the y axis.

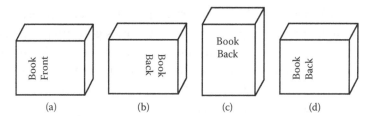

FIGURE 7.2 Position of the book after (a) a 90° rotation about the x axis and (b) a subsequent rotation of 180° about the z axis. In Figures (c) and (d), we show the results of the same operations, but performed in the reverse order.

The result depends on the order of the rotations, since Figure 7.2(b) does not coincide with Figure 7.2(d). Thus, multiplying element A of the group times element B does not give the same result as B × A. A group with this feature is called nonabelian.

The symmetries we discussed above involve a finite number of operations or group elements. If we consider a circle there is a symmetry with respect to rotations through any angle about the axis that goes through the center of the circle and which is perpendicular to the plane of the circle. The set of all such rotations is infinite and continuous. This corresponds to a continuous group with an infinite number of elements; again, one rotation (say by 10°) followed by another (say by 15°) corresponds to another element of the group (rotation by 25°).

Finally, we consider the set of all rotations in three-dimensional space. A sphere is invariant under all such rotations. The fundamental laws of physics are invariant under this symmetry. This is equivalent to saying there is no preferred direction in space as far as the laws of physics are concerned. It is important to emphasize that this concerns the laws that govern the universe, not the objects in the universe. If we consider the earth going around the sun and imagine the orbit to be a perfect circle, then there would be symmetry with respect to rotations about the axis perpendicular to the plane of the orbit but not to any other rotation (which would shift the plane of the earth's orbit). There is obviously a preferred axis, but we believe that this is the result of a chance occurrence when the solar system was formed and not the existence of any preferred axis in the fundamental laws.

One way to describe the invariance of physical laws is in terms of reference frames. You define the position of any point object in space by specifying the distance in three directions from some arbitrary origin, as in

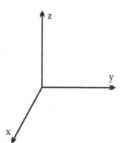

FIGURE 7.3   Reference frame for a three-dimensional space.

Figure 7.3. The motion of objects and the interactions that govern these motions are written in terms of a reference frame. The invariance with respect to rotation can be stated as the fact that the laws are not changed if you choose any other three perpendicular directions for your axes. There also is invariance with respect to the position of the origin of your reference frame; this is referred to as invariance with respect to space translation. The fundamental laws are the same in Tibet as they are in New York, and we believe they are the same in distant stars. Einstein extended this idea of laws that were independent of the reference frame to the case of moving reference frames, leading him to the theory of relativity.

There is a deep relation between symmetries and conservation laws. This was first pointed out by Emmy Noether, a mathematician born in 1882. For example, the symmetry under rotations in space leads to the conservation of angular momentum, and the symmetry with respect to space translation leads to conservation of linear momentum. There is also a symmetry relative to the translation of the time coordinate, and this is related to the conservation of energy.

7.2 MIRROR SYMMETRY: P FOR PARITY

Parity symmetry refers to the invariance under the reversal of all directions. It can be considered a change in the reference frame in which all the axes are reversed in direction. Given rotational symmetry, parity is equivalent to interchanging left and right, as shown in Figure 7.4. This is the same as a reflection in the mirror. The parity operation is indicated by the symbol P. The electromagnetic, strong, and gravitational interactions all are invariant under the parity transformation.

In classical physics, an elementary particle has no dimensions, and thus one cannot tell it from its mirror image. However, in the quantum world, where the particle is described by a field or a wave function, it is not quite

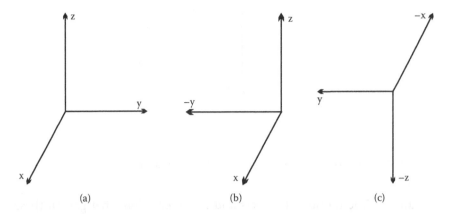

FIGURE 7.4 (a) Initial position of the reference frame. (b) Position obtained from (a) by performing a left-right symmetry. (c) Position obtained from (b) after a 180° rotation about the y axis. The result could be obtained directly from (a) through parity (i.e., through the reversal of all arrows across the origin).

that simple. Under parity the field or the wave function may transform into itself times –1; in this case we say that the particle has an intrinsic parity P = –1. Other particles may have P = +1. This distinction can play an important role in the physics of elementary particles. The electromagnetic interaction in quantum electrodynamics (QED) conserves parity provided the photon has P = –1. By convention, quarks, leptons, protons, and neutrons have P = +1. A feature of the Dirac theory of spin 1/2 articles is that antiparticles have opposite parities from particles, so that positrons and antiquarks have P = –1. Given an initial state of definite parity, the interactions that are invariant under parity require that the final state have the same parity; this is called parity conservation.

The parity of a system with several particles is obtained by multiplying all the parities involved; experts say that parity is a multiplicative quantum number. The ground state of a quark-antiquark system like the pion has parity P = –1 since, as noted, the antiquark has opposite parity from the quark. In the pion the quark and antiquark have opposite spins. There exists an excited state, called the rho-meson, in which the spins are parallel. Because quark and antiquark have spin 1/2, the total spin of the rho-meson is 1. Its parity is still P = –1. However, there are excited states in which the quark and the antiquark are rotating about each other with angular momentum, L. As first noted by Bohr, the angular momentum can only take values L = 0, or L = 1, or L = 2, and so on.* One can show that odd values of L correspond to parity P = –1, while even values of L correspond to parity P = +1. Thus, for a quark-antiquark system, P = –1 for even L and P = +1 for odd L.

An interesting example is the state made of a c quark and a c antiquark (ccbar) discussed in Chapter 5. The original discovery was of J/psi, which is the lowest ccbar state, having spin 1 and P = –1, analogous to the rho-meson. Other states were discovered later, in particular states with L = 1, which then had P = +1.

Another example is the hydrogen atom. Its ground state consisting of an electron and proton has P = +1. The first excited state, called the p-state, has L = 1, as originally described in Bohr's theory. This state has P = –1 and can decay to the ground state, emitting a photon while conserving parity, since the photon has P = –1. There is a further excited state with L = 2, which easily decays into the L = 1 p-state emitting a photon. (A transition from L = 2 to L = 0 is possible if the photon has an additional orbital angular momentum, but this is much less likely.)

Parity conservation has two implications. It means that particles have a definite parity, as discussed above. But it also means that a state with a definite parity always decays into final states that have the same parity. Consider a particle named X whose parity is unknown. Imagine that X decays via the strong or electromagnetic interaction into a final state f_+. The state f_+ is a shorthand notation for a final situation that may contain several particles with relative orbital angular momentum, but which, after all parities are taken into account, has parity +1. For example, we may have two pions in the final state with zero orbital angular momentum. The parity of each pion is –1, and the zero orbital angular momentum corresponds to parity +1. As a result, these two pions have parity –1 × –1 × +1 = +1. Because parity is conserved by the strong and electromagnetic interactions, if these interactions mediate the decay of X into f_+, then X must have positive parity. We may indicate this with a + subscript: X_+. Now consider another final state, f_-, whose parity is –1. An example is given by a final state with three pions and zero relative orbital angular momenta. Now, if parity is conserved, then X_+, which has parity +1, cannot decay also into f_-. To summarize, parity conservation implies that a particle cannot decay into two different final states of

---

* In units of the fundamental quantity of angular momentum, $\hbar$.

FIGURE 7.5 (a) Negative-helicity neutrino (or left-handed). The spin (in grey) points opposite to the momentum (in black). (b) Under parity, the spin stays the same and the momentum gets reversed. Thus, the spin has the same direction as the momentum, yielding a positive-helicity neutrino (or right-handed).

opposite parity. All decays mediated by the strong or electromagnetic interactions satisfy this constraint.

Because parity reverses directions, it reverses a particle's momentum. However, the spin direction of a particle as defined by Figure 3.1 is invariant under parity.[1] Helicity is the projection of a particle's spin in the direction of its motion. Given parity's action on spin and momentum, we conclude that the helicity of a particle also changes sign under parity. Under parity, spin 1/2 particles with negative helicity, such as the neutrino, will acquire positive helicity, and vice versa. This is illustrated in Figure 7.5.

The electromagnetic, strong, and gravitational interactions preserve parity. For a long time people believed that parity was a sacrosanct symmetry of nature. After all, why should left differ from right? Isn't it just a matter of convention? If you think it is merely a convention, you are not alone. Some of the most outstanding physicists of the first half of the twentieth century shared that opinion. They were baffled when experiments proved otherwise. Nature, through the weak interaction, does not conserve parity! One says that the weak interactions violate parity. In an August 1957 letter to the psychologist Jung (a former associate of Sigmund Freud), Pauli wrote, "Thus, it is now certain that 'God still is weakly left-handed' as I like to express it. ... Of such a possibility, I had not thought in the slightest before January of this year."[2]

Around 1956, people faced the following conundrum known as the tau-theta puzzle. A particle of a given mass decayed into the two-pion state we referred to above as f_+. And a particle of exactly the same mass decayed into the three-pion state we referred to above as f_-. Could these be two different particles with exactly the same mass? Or could parity be violated against all common wisdom? This problem was faced by two very young physicists, C. N. Yang and T. D. Lee. Together they braved where no others dared to go. They decided to look in the published literature for experiments that directly tested the possibility that parity might be violated in the weak interactions. They found none. Thus, they proposed a shocking solution to the tau-theta puzzle. Parity is violated by the weak interactions. As a result, one particle could decay through the weak interaction into f_+ and also into f_-.* They suggested several experimental tests of the hypothesis.

7.3 MADAME WU'S AMAZING DISCOVERY

Chien-Shiung Wu, a colleague of Lee's at Columbia University, led a group in the search for experimental evidence of parity violation in the decay chain

$$^{60}\text{Co} \rightarrow {}^{60}\text{Ni}^* \; e^- \; \bar{\nu}_e$$

The relevant features of this decay are the following. The cobalt nucleus has spin J = 5. Imagine that the spins are perfectly aligned† and that we measure how many electrons emerge making an angle theta (θ) with this direction. Figure 7.6a shows such a configuration, where the cobalt's spin is shown in grey and the electron's momentum in black. Under parity, the spin stays the same, but the electron's momentum reverses its sign. This is shown in Figure 7.6b. If parity were conserved, there should be as many

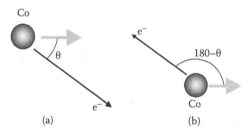

FIGURE 7.6 (a) Weak decay of cobalt nucleus (Co) with the electron emitted at an angle smaller than 90° with respect to the direction of Co's spin. (b) Parity symmetric figure. The Co's spin is in grey and the electron's (e⁻) momentum is in black.

---

* This particle is now known as the charged kaon, K⁺. Its parity is negative. This can be determined by analyzing the strong or electromagnetic interaction in which the K is produced. The decay into two pions violates parity, while the decay into three pions preserves parity.

† This was a difficult part of the experiment. To avoid misalignment due to thermal motion, temperatures of merely a hundredth of a degree above absolute zero (0.01 K) were required.

electrons appearing in the forward direction (making an angle with the spin smaller than 90°) as appear in the backward direction (making an angle with the spin larger than 90°). In 1957 Wu's group found an excess of electrons in the backward direction. That is, the situation depicted in Figure 7.6b prevails.

A number of experiments followed almost immediately. Garwin, Lederman, and Weinrich, on the one hand, and Friedman and Telegdi, on the other, looked into $\pi^+ \to \mu^+ \nu_\mu$ followed by $\mu^+ \to e^+ \bar{\nu}_\mu \, \nu_e$. The experiments may be more or less complicated, but all confirm that parity is violated by the weak interactions.

The first decay is particularly interesting. The pion has spin zero, while the antimuon and the neutrino have spin 1/2. Due to conservation of angular momentum, the spins of the products of the pion decay must be opposite of each other (thus adding to spin zero). One possibility is shown in Figure 7.7a.

The other possibility, shown in Figure 7.7b, is the parity transformation of Figure 7.7a. The relevant difference between the two figures is not that the antimuon of the first figure is moving to the left while the muon of the second figure is moving to the right. That can be altered by a simple rotation, leading to Figure 7.7c. Since physics is invariant under rotations, this can still be viewed as the parity transformation of Figure 7.7a. The relevant difference between Figure 7.7a and its parity transformed, whether

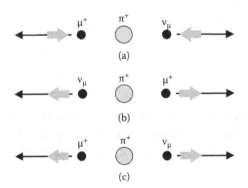

FIGURE 7.7 (a) Weak decay of the charged pion in a reference frame where the pion is at rest. (b) Parity symmetric figure, with the momenta reversed (the antimuon, which was going to the left, is now going to the right). (c) Rotation of (b) by 180°. This may still be taken as the parity transformation of (a). The spins are shown in grey, the momenta in black. The parent pion is shown in grey because it will disappear, with the antimuon and the neutrino appearing in its place.

it is represented by Figure 7.7b or by Figure 7.7c, is that the helicity of the antimuon gets reversed—and likewise for the neutrino. The neutrino and antimuon have negative helicity in Figure 7.7a, while they have positive helicity under a parity transformation.

If parity were conserved, both possibilities should be observed with exactly the same rate. The result could not be more striking. The situation in Figure 7.7a is readily observed; the situation in Figure 7.7c has never been seen! Nature chose to violate parity to the maximal degree. One situation occurs; its parity transformation does not!

The results discussed so far led several physicists to propose that for neutrinos, which were presumed to be massless, the weak interaction only involved positive-helicity antineutrinos and negative-helicity neutrinos. One cannot make the same assumption for particles with mass because positive helicity changes to negative helicity if you run past the particle faster than the particle is moving. For massive particles a new concept, called chirality, is introduced. The weak interaction involves only right-handed (R) chiral antifermions and only left-handed (L) chiral fermions. This means that if the emitted fermion has an energy much larger than mc^2 its (R,L) chirality implies mostly (+,–) helicity. But there exists a probability proportional to m^2 that the (R,L) fermion will have (–,+) helicity.

So, physicists conjectured that the weak interactions driving $\pi^+ \to \mu^+$ ν_μ involve only chiral left-handed fermions and chiral right-handed antifermions. This information tells us something else about the decay in Figure 7.7a. The weak interaction produces a chiral left-handed neutrino and a chiral right-handed antimuon. If their masses were zero, this would correspond to a negative-helicity neutrino, as in Figure 7.7a, and a positive-helicity antimuon. But Figure 7.7a has an antimuon with negative helicity. Thus, if the mass of the muon were zero, the process in Figure 7.7a would not take place either. That is, all helicity configurations of the decay $\pi^+ \to \mu^+$ ν_μ would be forbidden if the muon and neutrino were massless. The decay would not take place. So, the process in Figure 7.7a only occurs because the antimuon has some mass. And the chiral right-handed antimuon can be seen as a negative-helicity antimuon with an amplitude proportional to its mass. Because the probability is the square of the amplitude, the rate for the decay $\pi^+ \to \mu^+$ ν_μ is proportional to the square of the muon mass.*

---

* In Part C evidence is presented that the neutrinos do have a mass, but this mass is much too small to be relevant here.

Now consider the decay $\pi^+ \to e^+ \, \nu_e$. The same argument means that its rate should be proportional to the square of the electron mass. With very simple helicity arguments we found a striking prediction. The rate of the decay $\pi^+ \to e^+ \, \nu_e$ must be much smaller than the rate for $\pi^+ \to \mu^+ \, \nu_\mu$. The ratio between the two should be of the order of the electron mass squared divided by the muon mass squared.[3] Experiments verified such a large difference in rates.

By mid-1957, parity violation became firmly established, taking the physics community by storm. Pais recalls: "On the subject of weak interactions alone, 1000 experimental and 3500 theoretical articles as well as 100 reviews (all in round numbers) appeared between 1950 and 1972, written by authors from 50 countries."[4] For this unexpected discovery, Lee and Yang received the 1957 Nobel Prize in Physics, one of the fastest recognitions ever. Madame Wu did not!

7.4 C FOR CHARGE CONJUGATION

Charge conjugation is a transformation, usually represented by C, which exchanges all particles by the corresponding antiparticles, and vice versa. The prediction of antiparticles appeared from Dirac's effort to combine quantum mechanics with relativity. So, charge conjugation is a quantum symmetry with no classical analogue. Like parity, charge conjugation is a good symmetry of the strong, electromagnetic, and gravitational interactions. We say that these interactions preserve C.

Charge conjugation exchanges particles with antiparticles. One knows that conserved quantities like charge and baryon number have their signs reversed with respect to the corresponding particle: This has two consequences. First, we cannot attribute a charge conjugation quantum number to a charged particle. That can only be made for particles with no electric charge (neutral particles), no baryon number, etc... That is, the word "color" gets substituted by "baryon number." Second, charged particles are distinct from their antiparticles, but neutral particles may or may not coincide with their antiparticles.

The photon and the neutral pion (π^0) coincide with their antiparticles. Electromagnetic fields are generated by moving charges, which change sign after charge conjugation. As a result, the photon has C = –1. Like P, C is a multiplicative quantum number. Since π^0 decays electromagnetically into two photons (and not three), it has C = –1 × –1 = +1.

But there are neutral particles that are different from their antiparticles. For example, there are two neutral kaons, K0 and K0bar (pronounced

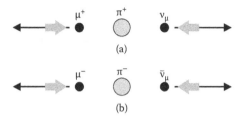

FIGURE 7.8   (a) Weak decay of the positively charged pion in a reference frame where the pion is at rest. (b) Charge-conjugated decay.

"K-zero" and "K-zero-bar"), which are the antiparticles of each other. This can be easily recognized by considering their quark components. K0 is made from sbar (s antiquark) and d, while K0bar is made from s and dbar:

$$K^0 = \bar{s}\, d, \qquad \bar{K}^0 = s\, \bar{d}$$

One can attribute an additive quantum number $S = -1$ to the s quark and $S = +1$ to the s antiquark. The difference in this quantum number explains why K0 and K0bar cannot be their own antiparticles. Charge conjugation exchanges K0 with K0bar.

Lee, Oehme, and Yang pointed out that Wu's cobalt experiment also indicated C violation by the weak interaction. This violation is more striking in the decay $\pi^+ \rightarrow \mu^+ \, \nu_\mu$. Figure 7.8a shows the configuration introduced in Figure 7.7a. Figure 7.8b shows its charge conjugate. The situation in Figure 7.8a is readily observed; the situation in Figure 7.7b has never been seen! Nature chose to violate charge conjugation to the maximal degree. One situation occurs; its charge conjugate does not!

7.5 CP SYMMETRY

Shortly after Wu's discovery of P and C violation, several scientists discussed the possibility that the weak interactions might preserve the combined CP symmetry. By this one means the following. Some given process is studied. Then one makes a parity transformation on that process, followed by a charge conjugation. This is dubbed the CP-transformed or CP-conjugated process. If CP is conserved, then the original process and its CP-conjugated one should have the same rate. An example is shown in Figure 7.9. Choosing to transform under C and then under P, or vice versa is immaterial; either way, one ends up at the same CP-conjugated process. The top left illustration shows a positively charged pion decaying into a

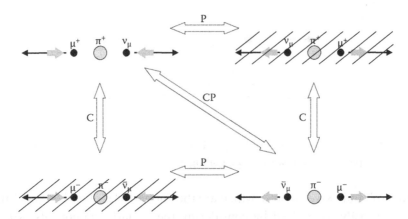

FIGURE 7.9 Top left: Weak decay of the positively charged pion in a reference frame where the pion is at rest, with a negative-helicity neutrino in the final state. Also shown are the decays related to it by P, C, and CP. Those that have never been found are slashed.

negative-helicity antimuon and a negative-helicity neutrino. This decay has been observed. The P-transformed decay includes a positive-helicity neutrino, while the C-transformed decay includes a negative-helicity antineutrino. As mentioned before, these decays have never been seen, so we have crossed them out.* But the CP-conjugated decay has been seen. And, within experimental errors, these π^+ and π^- decays have the same rates. Thus, CP seemed a good symmetry.†

It was then shown by Feynman and Gell-Mann and by Marshak and Sudarshan that there was a simple way to modify the standard weak interaction theory originally due to Fermi to allow for C and P violation but to maintain CP invariance. Each field in the weak interaction was replaced by a left-handed chiral field.‡ This came to be known as the V – A theory (pronounced "V minus A theory") and was successful in explaining all weak interaction phenomena until 1973, with one small exception.

The exception was the discovery in 1964 that CP was not an exact symmetry. In studying the decays of neutral kaons, Jim Cronin, Val Fitch, and

---

* In fact, no one has ever made an unequivocal identification of a positive-helicity neutrino or a negative-helicity antineutrino. We will come back to this point when we discuss neutrino masses.
† The equality of π^+ and π^- lifetimes is tested to one part in a thousand. But this is, in fact, a test of CPT invariance.
‡ For the massless neutrino this meant the neutrino had negative (left-handed) helicity and the antineutrino positive (right-handed) helicity. For the electron and other massive particles it meant that the interaction produced a left-handed particle in the limit of zero mass.

collaborators found a small violation of the CP symmetry. There was no doubt that the symmetry was violated, but CP violation was not seen in any other system for more than 30 years. The search for CP violation is now one of the most exciting endeavors in the study of elementary particles and is the subject of Part B of this book.

7.6 T FOR TIME REVERSAL

Fundamental laws, from Newton's laws to quantum electrodynamics, can be shown to be invariant under the reversal of time. Under time reversal, velocities, momenta, and angular momenta all reverse their direction. Also, the initial and final states of any process are reversed. An elementary example is a particle starting at rest and gaining velocity as it falls to the earth. Neglecting friction, the time reverse process is one in which the particle starts with a high velocity moving away from the earth and loses velocity until it comes to rest. In particle physics it is possible to test time reversal invariance by relating the rate at which two particles a and b collide to produce particles c and d to the reverse reaction in which c and d collide to produce a and b. This relation is known as the principle of detailed balance and has been verified with considerable precision in certain nuclear reactions.

It sometimes seems strange to talk about time reversal when we know that complex processes cannot be reversed. If you drop an egg on the floor and it breaks apart, you do not expect to ever see the reverse, in which the egg reconstructs itself and rises back up. For complex processes there is indeed an "arrow of time," and this is encoded by physics in the second law of thermodynamics, which concerns a quantity called entropy, a measure of disorder, that always increases with time.

However, physicists have also found that the microscopic laws describing those phenomena (except, as we will see, the weak interaction) remain the same if we reverse the direction of time. There seems to be a contradiction. How can the macroscopic irreversibility arise from microscopic laws that are time reversible? Lee has come up with an analogy that might help.[5] Consider one thousand drivers departing from New York's Kennedy Airport and moving 1,000 km, each in its own direction. Each driver passes many intersections and, at each intersection, decides at random which path to follow. After the drivers stop, they will be distributed all over the map, within a circle with a 1,000 km radius. Now imagine that they reverse directions and again travel 1,000 km using the same rule. At each intersection, each driver decides at random which direction to follow. You know intuitively that it will be unbelievably unlikely to find all cars back at Kennedy Airport. Notice that the same rule was used. But having the cars scatter all over the map is much more likely than having them all meet by chance at the same location. The greater the number of cars and intersections, the greater will be the likelihood difference. For the typical 10^{23} particles in some portion of matter with countless collisions, one time direction completely overwhelms the other. So, events follow the line of increased likelihood. Entropy, which is a measure of how likely a given situation actually is, tends to increase.

In the early 1950s, a number of theoretical physicists wrote down all the possible interactions they could think of in order to probe the possibilities of new physics. The interactions were all written in the language of relativistic quantum field theory, like QED, with particles interacting at a point in space, like the electron interacts with the photon in QED. They found it was easy to have theories that violated C and P and T, but that no theory violated the product C × P × T. This led to a theorem that all reasonable theories were invariant under CPT. As a result, any theory that explains CP violation will also violate T invariance. So far it has not been possible to directly observe T violation because CP violation has been observed only in decay processes, and it is not possible to reverse a decay. It still remains important to test the CPT symmetry experimentally. The best tests involve the equality of the masses of particles and antiparticles, while other tests probe whether their charges and magnetic moments are equal in magnitude and opposite in sign. So far all tests are consistent with CPT invariance.

NOTES

1. Note that the direction of spin (or angular momentum) given by the long grey arrow in Figure 3.1 depends on both the direction of velocity of the particle (given by the short black arrow) and the direction of the particle (black dot) from the center of the circle. Technically, the direction of the angular momentum is given by the cross product of the momentum vector, $\vec{p}$, and the vector representing the position of the particle from the centre $\vec{r}$: $\vec{L} = \vec{r} \times \vec{p}$. Thus, reversing both of these vectors does not change the direction of the angular momentum. One says that the orbital angular momentum and the spin are pseudovectors. They rotate like ordinary three-dimensional vectors. But under parity, ordinary vectors reverse their direction, while pseudovectors stay the same.

2. C. P. Enz, *No Time to Be Brief: A Scientific Biography of Wolfgang Pauli* (Oxford: Oxford University Press, 2002), p. 519.
3. This would give 2.3×10^{-5}. But this is not the whole story. The fact that the electron has a much smaller mass than the muon gives it a slight compensating factor (known as phase space). Combining helicity and phase space effects, one finds

$$\frac{Rate[\pi^+ \to e^+ \nu_e]}{Rate[\pi^+ \to \mu^+ \nu_\mu]} = \left(\frac{m_e}{m_\mu}\right)^2 \left(\frac{m_\pi^2 - m_e^2}{m_\pi^2 - m_\mu^2}\right)^2 = 1.2 \times 10^{-4}$$

which agrees with experiment.
4. Abraham Pais, *Inward Bound* (Oxford: Oxford University Press, 2002), p. 533.
5. T. D. Lee, *Particle Physics and Introduction to Field Theory* (Chur, Switzerland: Harwood Academic Publishers, 1990).

CHAPTER 8

Emergence of the Standard Model

FERMI ALWAYS SAID THAT his weak interaction theory of 1937 was inspired by quantum electrodynamics (QED). In QED, the basic interaction involves an electron going into an electron plus a photon; Fermi had a neutron going into a proton, with the emission of an electron and an antineutrino. Thus, the combination of the electron with the antineutrino played the role of the photon. Also in the 1930s, a different analogy with QED was suggested by Felix Klein. Consider the diagram in Figure 8.1a, in which an electron near a nucleus produces a positron-electron pair due to the exchange of a virtual photon. The analogue is Figure 8.1b, where the virtual photon is replaced by a new particle, a charged boson W⁻. Then, the fundamental interaction, like in QED, involves a vertex with just three particles: fermion-fermion-boson.

This idea became very interesting in the late 1940s with the great success of QED, when Feynman and Schwinger showed that QED was renormalizable, allowing the calculation of higher-order effects (corresponding to complicated Feynman diagrams), like a small correction to the magnetic moment of the electron that agreed perfectly with the experiment. The key to renormalization was a property of QED called gauge invariance. It was another 20 years until the standard model of weak interactions emerged based on a gauge theory.

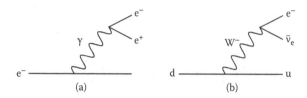

FIGURE 8.1   Analogy between (a) in which an electron-positron pair is produced via a virtual photon and (b) that shows the beta decay interaction of a d quark going into a u quark, an electron, and an antineutrino, due to exchange of a virtual W⁻ boson in analogy to (a).

8.1 WEINBERG: A MODEL OF LEPTONS

The essential quantum concept discussed in Chapter 2 was that electrons, photons, etc., which we may think of as particles, propagate like waves. The question arises as to what is "waving": What is it that varies in time and space like a wave? Such a question arose in a different context when it was shown by Young in the early nineteenth century that light was a wave. The answer given then was that all space was filled with a mysterious substance called the ether, and it was the ether that was oscillating like the water in a water wave. After the work of Maxwell and Einstein it became clear that there was no ether; what varied in time and space were the electric and magnetic fields.

Wave particles are specified by a quantum field: the symbol Ψ for the electron and A for the photon. These fields are specified by a complex number at every point in space and time. A remarkable feature of the electromagnetic interaction is that there exist a whole set of possibilities for Ψ and A that give the same results. This is what is called gauge invariance.* Another way of saying this is that by insisting on this local gauge invariance, you can predict the exact form of the electron-photon interaction. Of course, historically the interaction form came from classical electrodynamics. On the other hand, the idea that you could use gauge invariance to predict the form of interactions became the inspiration for the standard model.

In 1967 a short paper, only two and a half pages long, entitled "A Model of Leptons," appeared in the journal *Physical Review Letters*. For the next 3 years there were essentially no references to this paper. It seemed to be an arbitrary new theory of weak interactions that didn't even solve the

---

* Complex numbers and gauge invariance are discussed in Appendix 3.

one experimental problem of the old theory, which was CP violation. In 1979 Steven Weinberg shared the Nobel Prize in Physics in honor of this paper. Weinberg's paper had presented what is now the standard model of weak interactions. Weinberg restricted his original paper to leptons: electrons, muons, and their neutrinos. In 1970 Glashow, Iliopoulos, and Maiani (GIM) showed that any reasonable weak interaction theory needed a fourth quark, the charm quark discovered in 1975 (see Chapter 9). The GIM paper had no reference to Weinberg's 1967 paper, but shortly thereafter Weinberg used the GIM idea to extend his theory to quarks.

The hope of Weinberg was that his theory might be renormalizable. In 1971 a young Dutch student, Gerard 't Hooft, working with Martinus Veltman showed that this was true: they won the Nobel Prize in 1999. As a result of this paper the new theory began to excite the physics community.

The starting point of Weinberg was local gauge invariance. In QED, the U(1) gauge invariance involved a single gauge boson, the photon. In order to extend this to allow for W^+ and W^- bosons, it had been shown in 1954 by C. N. Yang and Robert Mills that one required three intermediate bosons: W^+, W^-, and an electrically neutral W^0. This was called SU(2) gauge invariance. However, the gauge invariance required the gauge bosons to be massless, like the photon. This leads to an interaction that falls inversely as the square of the distance, while it was known that the weak interaction had a very short range in order to explain the energy spectrum in weak decays.

Yukawa had shown (see Chapter 5) how to make interactions with virtual intermediate particles fall off rapidly with distance: give the intermediate boson a mass. This violated gauge invariance. Analysis of the electron spectrum in the decay of the muon required a mass of the W boson several times that of the proton. In 1963 there was an experimenter who thought he had discovered a W boson with a mass three times that of the proton. When the W was finally discovered, as discussed below, it turned out to have a mass eighty-five times that of the proton!

Weinberg's proposal starts with a theory that is gauge invariant and with massless W bosons. Then, following an idea due to Peter Higgs, he introduces into the theory a new field, now called the Higgs field, which has a weak charge allowing a gauge-invariant interaction with the W bosons. Then comes the Higgs trick. Ordinarily, if you have a field like the photon field A in the vacuum with zero photons, the value of A is zero. However, the Higgs field is configured to have a nonzero value, v, in the vacuum; this is referred to as a vacuum expectation value (vev). There is now an interaction term of the W with this vev, and it is equivalent to a mass for the W bosons; this destroys the gauge invariance. This mechanism is called spontaneous violation of gauge invariance. It is the fact that the violation of gauge invariance is entirely due to this mechanism that renders the theory renormalizable.

The theory is not quite this simple. It is necessary to go back and look at QED and avoid giving mass to the photon. The solution was to add to the W interactions an interaction like QED, which had a U(1) gauge invariance with a field B^0 in place of the photon. The combination of the two interactions was called an SU(2) × U(1) gauge theory. There were then two coupling constants, g and g' (pronounced "g-prime"). It was then found that by mixing the W^0 and the B^0, it was possible to produce a massless particle, the photon, and reproduce QED. The other combination (mainly W^0 with some B^0 mixed in) would be heavier than the W bosons. This combination was called Z^0.[1] The basic idea for this mixture is represented in Figure 8.2.

The theory provided an important relation between the weak and electromagnetic interactions. For this reason, it is often referred to as the electroweak theory. The weak coupling constant, g, measures the strength of the electron–neutrino–W vertex, analogous to the charge, e, of the electron appearing in the vertex with the photon. The theory actually required that g was about twice the electron charge. This sounds very strange: the weak interaction is stronger than the electromagnetic? The explanation is that the W boson has a very large mass: at low energies it is very difficult to produce a virtual particle with such a large mass, and so the interaction

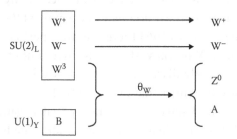

FIGURE 8.2   Relation between the original gauge bosons of SU(2)$_L$ × U(1)$_Y$ and the physical gauge bosons. There remains the U(1) gauge symmetry of electromagnetism with A representing the massless photon. The photon and Z^0 are combinations of the neutral gauge bosons in the original theory, involving the Weinberg angle, θ_W.

is highly suppressed. At sufficiently high energies, measurements confirm that the weak interaction is stronger than the electromagnetic.

There was still another problem, one that followed from parity violation. Only the left-handed leptons, like the left-handed electron field, e_L, had weak interactions and a weak charge. The right-handed electron field, e_R, had no weak charge. This required that the original gauge-invariant theory have massless leptons, since a left-handed particle with mass looks like a right-handed particle if you move by it fast enough. Amazingly, it turned out that the Higgs trick could also solve this problem. You include in the original massless theory an interaction between the Higgs field, the e_L field, and the e_R field. Since the Higgs field has a weak charge, this interaction is gauge invariant. The strength of the interaction is given by a number Y_e and then, because of the vacuum expectation value of the Higgs field, v, the electron gets a mass given by $Y_e \times v$.* The charged leptons and the quarks get mass in the same way; the values of the masses depend on a set of arbitrary parameters like Y_e, known as Yukawa couplings. In Weinberg's theory there is no right-handed neutrino and the neutrino is massless. The question of neutrino mass is the subject of Part C.

What the theory says is that there is a Higgs field everywhere in space, and this gives particles their masses. If this seems mysterious to you, you are not alone. Nobel Prize-winning physicist Leon Lederman wrote a book about the Higgs particle and gave it the name *The God Particle*, since it seemed that the Higgs was responsible for all the mass in the universe. However, once you write the theory down with all the particles having their masses, there are no observable consequences of the field in the vacuum. What is an observable and an essential feature of the theory is the excitations of the field, which are massive spinless neutral particles called Higgs bosons. The search for the Higgs boson is described in Part D.

8.2 THE EXPERIMENTAL TRIUMPH OF THE STANDARD MODEL

A fundamental prediction of the theory is a new type of weak interaction due to the exchange of the Z^0 boson. Interactions involving this boson are known as neutral current interactions; examples are illustrated in the diagrams of Figure 8.3. A neutrino could scatter off a neutron or proton or electron and remain a neutrino. In 1973 a group at CERN in Geneva announced that they had detected neutral current events. A beam of

---

* See Part D for explanations of the vacuum expectation value.

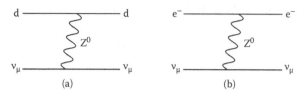

FIGURE 8.3 Feynman diagrams for neutral current interactions involving exchange of the Z boson.

muon type neutrinos emerged from the decay of pions produced by the collision of the proton beam with nuclei. This beam then entered a large bubble chamber with the fanciful name Gargamelle containing a heavy liquid. The neutral current scattering was detected by the observation of a neutron track originating in the middle of the chamber with nothing seen coming in and nothing seen coming out. The two "nothings" were the incoming and outgoing neutrinos. There was also one beautiful event with an electron appearing, indicating the scattering of the neutrino from an electron.

After this announcement, there was much discussion whether neutral currents had really been observed. In the fall of 1973 word came from Carlo Rubbia, who was carrying out an experiment at Fermilab, that he could not find neutral current events. As more data were acquired, the conclusion became clear, and Rubbia also found strong evidence for neutral currents.

There was also in the theory a neutral current interaction between the electron and the neutron. This was much harder to detect because the low-energy electrons available would scatter from a nucleus due to the much stronger electromagnetic interaction; the poor weak interaction would be overwhelmed. The solution was to make use of the fact that the weak interaction should violate parity conservation. In the experiment carried out at Stanford in 1978 by Charles Prescott and collaborators, a beam of polarized electrons was elastically scattered from deuterons (the deuteron is the nucleus of heavy hydrogen containing one neutron and one proton). The parity violation was detected by measuring the difference in the scattering when the electrons had negative helicity (left-handed polarization) from when they had positive helicity (right-handed polarization). A difference of 1 part in 10,000 was found, in complete agreement with the theory.

There remained the crucial test of the theory: the actual detection of the W and Z particles. However, the W was extremely heavy and the Z even heavier; the beams from the highest energy accelerators in existence, at Fermilab and CERN, did not have enough energy to produce them. Various proposals for new, more powerful accelerators were advanced. Among these was a novel idea, originally due to Peter McIntyre, a 27-year-old American physicist: a high-energy beam of protons would be collided against an equally high energy beam of antiprotons. Out of the collision would come many particles; occasionally they would include a W or a Z.

Carlo Rubbia became very excited about this proposal and urged the director of Fermilab to work on it. When Fermilab made no decision, Rubbia returned to CERN and in January 1978 it was decided to construct such an accelerator in Geneva. The most difficult aspect of the machine was somehow to collect the antiprotons to be accelerated. They first had to be produced by means of the collision of a relatively low energy proton beam with a nucleus, and then they had to be streamlined or cooled before they could be injected into the big accelerator ring. The way to do this was developed by a Dutch engineer, Simon van der Meer. Meanwhile, Rubbia took charge of the construction of a huge detector named UA1. There was also a smaller detector named UA2 to be placed at a different place around the accelerator ring. In July 1981 the first collisions of protons with anti-protons were observed.

In January 1983, Rubbia announced that the W particle had been detected in UA1. Almost immediately the result was confirmed by the leader of UA2. Because the W decays immediately after being produced, it is the products of the decay that must be detected. In this case, it was the electron from the decay of the W into an electron and a neutrino. While the neutrino left no track, the clue to its presence was missing energy, the same clue that led Pauli to invent it in the first place. In June 1983, Rubbia announced the detection of the Z^0 decay. In 1984 Rubbia and van der Meer shared the Nobel Prize in Physics. The new standard model of weak interactions was firmly established.

NOTE

1. The fields are related by $A = B \cos\theta_W + W_3 \sin\theta_W$, for the photon, and $Z = - B \sin\theta_W + W_3 \cos\theta_W$, for the Z boson.

CHAPTER 9

Flavor Physics

WEINBERG'S ORIGINAL PROPOSAL FOR the standard model was limited to leptons. There were two pairs of leptons: (electron, electron neutrino) and (muon, muon neutrino). At the time, the hadrons consisted of three quarks: u, d, and s.

We now know that the standard model has three families of fermions. In each family, there is one charged lepton, one neutral lepton, three quarks with charge –1/3 (one for each color), and three quarks with charge +2/3. Most times we ignore the color repetition of the quarks, referring to only one charge –1/3 quark and one charge +2/3 quark per family. Since there are three families, we get a total of six quarks: down, up, strange, charm, bottom, and top. These are known as the quark flavors. The word *flavor* is sometimes used to refer to the various leptons. For example, the electron, the muon, and the tau would be the three possible flavors of charged leptons.

In contrast to the gauge sector, where the gauge structure and a few gauge couplings determine all interactions, the interactions of the quarks and leptons with the Higgs field (known as Yukawa couplings) are completely arbitrary. Moreover, we have no explanation of why there are only three families. These are considered by most the ugliest theoretical features of the standard model. But, by the same token, there is much that can be studied experimentally. The study of the Yukawa couplings and their experimental consequences goes under the name of flavor physics.

9.1 STANDARD MODEL WITH TWO FAMILIES

In the early 1960s people still described weak decays through Fermi's four-fermion interaction, though some small discrepancies between neutron and muon decays were creeping up. The same theory explained the decay of the charged pion into an antimuon and a muon neutrino. So, when strange particles appeared, Fermi's theory was called upon once more. The idea is shown in Figure 9.1, in terms of today's quark picture. Assuming the same interaction strength for the quark combinations us and ud, Fermi's theory would imply a certain relation between the kaon and the pion decays into an antimuon and a muon neutrino. The relation did not hold, and some modification was needed. The same problem appeared in the relation between the decays of baryons with and without strangeness.

In 1963, Nicola Cabibbo solved all discrepancies with one single parameter. In today's quark picture, Cabibbo's solution means that the coupling of the W to the u and s quarks (denoted by g_{us}) is different from its coupling to the u and d quarks (g_{ud}). Cabibbo wanted to keep some relation between these couplings and the coupling of the W to the muon and muon antineutrino (g), in an argument known as universality. This he did through the introduction of an angle θ_C, now known as the Cabibbo angle. The technical solution was to relate these couplings in the same way we relate the sides of a triangle.[1] This is shown in Figure 9.2. The current value for the angle is around 13°. Cabibbo's analysis is impressive, especially if you remember that it was put forth well before quarks were established.

This has implications for the couplings of the quarks with the $SU(2)_L$ gauge bosons. It implies that while the muon and the muon neutrino are organized into a doublet, the up quark should appear in a doublet with a

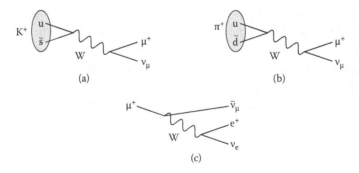

FIGURE 9.1 Feynman diagrams for the transitions: (a) $K^+ \to \mu^+ \nu_\mu$, (b) $\pi^+ \to \mu^+ \nu_\mu$, and (c) $\mu^+ \to \bar{\nu}_\mu\, e^+ \nu_e$.

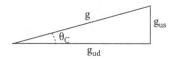

FIGURE 9.2   Cabibbo's solution: The couplings g_{us}, g_{ud}, and g are related like the sides of the triangle.

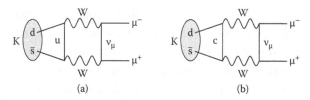

FIGURE 9.3   Contribution of the quark's interaction with Ws to the decay of kaons into muon and antimuon. These are known as box diagrams. Contribution from: (a) the up quark and (b) the charm quark.

combination of s and d quarks. And the combination is controlled by the Cabibbo angle:

$$\begin{pmatrix} \text{neutrino} \\ \text{electron} \end{pmatrix}_L, \quad \begin{pmatrix} u \\ d_W \end{pmatrix}_L$$

The index W is a reminder that this field (which couples with the u quark to the W boson, with strength g) is part d quark, part s quark.[2]

In 1970 Glashow, Iliopoulos, and Maiani realized that this could not be all. If d_W coupled in this way with the $SU(2)_L$ gauge bosons, it would also couple to the Z^0, implying that an s quark could turn into a d quark by emitting a Z^0. This possibility is known as flavor-changing neutral currents, because the neutral Z^0 would permit a change between two quark flavors. But, because the Z^0 also couples to muon and antimuon, this implied a decay of kaons into muons and antimuons with a rate much higher than allowed by experiment. If that were not enough, the coupling of u through the W to d and s would also give too large a contribution. This is shown in Figure 9.3a. The solution was to introduce a fourth quark named charm, coupling to a different combination of the d and s quark, dubbed s_W:[3]

$$\begin{pmatrix} c \\ s_W \end{pmatrix}_L$$

This solved both problems. First, the combined couplings of d_W and s_W to Z^0 only allow it to couple s quark with s antiquark or d quark with d antiquark, forbidding all flavor-changing neutral currents. Second, the presence of a charm quark would allow the contribution from Figure 9.3b to partly cancel the diagram contribution from Figure 9.3a.[4] The cancellation would be exact if the mass of the charm were exactly equal to the mass of the up quark, and it would not be enough if the mass of the charm were too large. This limited the mass of the then proposed charm quark to lie between 1 and 3 GeV. As discussed before, the discovery of charm in 1974, together with correct predictions for J/ψ's excited states, marked a major success for the theory and provided strong evidence for the quark model. The analysis of the data required a charm quark mass in the neighborhood of 1.5 GeV.

At the end of 1974, the standard model had two families of leptons and two families of quarks. The left-handed fields are organized in doublets as

$$\text{First family:} \quad \begin{pmatrix} \text{electron-neutrino} \\ \text{electron} \end{pmatrix}_L, \begin{pmatrix} u \\ d_W \end{pmatrix}_L$$

$$\text{Second family:} \quad \begin{pmatrix} \text{muon-neutrino} \\ \text{muon} \end{pmatrix}_L, \begin{pmatrix} c \\ s_W \end{pmatrix}_L$$

The physical quarks are not d_W and s_W but rather d and s. The two sets are related by a rotation, as shown schematically in Figure 9.4.[5] Unfortunately, the standard model with two families cannot explain CP violation. This was known even before the discovery of the charm quark in 1974. In a

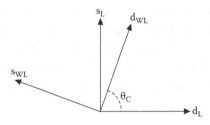

FIGURE 9.4   Schematic representation of the change from d_W and s_W to the physical d and s.

twist of history, what we now believe to be the solution to this problem was already hidden in one paragraph of a paper by Kobayashi and Maskawa published in 1973. Among other possibilities, they pointed out that a third pair of quarks would provide a method of allowing CP violation in the standard model.

9.2 STANDARD MODEL WITH THREE FAMILIES

In 1975 Martin Perl, working at the Stanford Linear Accelerator Center (SLAC), announced the discovery of a third charged lepton called the tau, seventeen times heavier than the muon and twice as heavy as the proton.* Perl's experiment involved the same accelerator that had discovered charmonium the year before. The evidence for the tau was a set of events in which the electron and positron collided and the final state contained one muon and one positron and nothing else (or one antimuon and one electron and nothing else). This was interpreted as a final state containing a tau and an antitau in which one of the taus immediately decayed into a muon and neutrinos, and the other into an electron and neutrinos, as shown in Figure 9.5. The tau lifetime was so short you could not detect any track of the tau itself, only the decay products.

The mass of the tau was just about equal to the expected mass of the charm quark. When Perl presented his results, many people thought they had to involve the decay products of a particle containing a c quark. However, as Perl's data came in, such an explanation didn't work. Could it be that the similarity in mass of the tau and the charm quark was just a coincidence? The answer is yes, coincidences do happen!

Over the years since 1975, many decay modes of the tau-lepton have been discovered. Each of them involves at least one neutrino, and this

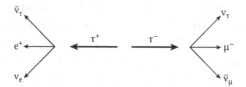

FIGURE 9.5   Schematic representation of one possible tau-antitau decay.

---

* The names hadron and lepton were derived from the Greek to indicate that hadrons were heavy and leptons light. With the discovery of the tau this was no longer true, but the names remained.

was presumed to be a third type of neutrino: the tau neutrino. A variety of pieces of evidence confirmed that there were three types of neutrinos. However, direct detection of the tau neutrino is very difficult, since one cannot create a beam of tau neutrinos like one can for electron neutrinos or muon neutrinos. On the rare occasion that a tau neutrino, coming from a tau decay, interacts, it will again produce a tau. And that tau will again decay, meaning that most detectors will fail to see a distinctive track arising from the interaction of a tau neutrino. Still, it is possible to see the track in special sensitive emulsion detectors, and in that way a few tau neutrinos were finally directly detected in 2000 at Fermilab's DONUT experiment.

The discovery of a third pair of leptons led many to think about a third pair of quarks. In the fall of 1975 the E288 experiment started at Fermilab, looking for quarks heavier than the charm. The spokesman was Leon Lederman, already a famed experimentalist having, among others, discovered the muon neutrino for which he would eventually receive the 1988 Nobel Prize with Schwartz and Steinberger. At some point, a signal was found around 6 GeV that could be the searched-for bottom-antibottom bound state. The particle was named upsilon, after the Greek letter U. This turned out not to be a real particle but rather a fluctuation of the data, hence known as "oopsLeon" in honor of the spokesman. The real particle was found by Lederman's group in 1977 at 9.5 GeV. Close by were its excited states. Because of the high energy involved, the findings were a perfect match to the quantum chromodynamics (QCD)-based perturbative calculations, assuming a bottom quark around 4.5 GeV. This was the first of a series of landmark discoveries at Fermilab.

As with the charm, once the bottom quark of charge −1/3 was found, consistency of the standard model and detailed calculations involving diagrams similar to the ones in Figure 9.3 required a further quark of charge +2/3, named top. Top quarks turn out to be extremely heavy; they are around ten thousand times heavier than the up quark. They exist for such a short time that, unlike all lighter quarks, they do not form a top-antitop bound state. A top quark is detected through its known decay products. One possibility is shown in Figure 9.6. The initial proton and antiproton collide, resulting in a top-antitop pair. The top quark decays into a W^+ gauge boson and a bottom quark, and similarly for the antitop quark. One W decays into a muon and a neutrino, and the other into quarks that form two jets. The bottom and antibottom eventually decay, forming two further jets. Since the neutrino goes undetected, this is known as a "lepton plus four jets" event.

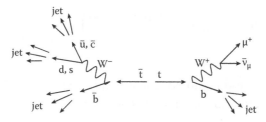

FIGURE 9.6 Schematic representation of one possible top-antitop decay as a "lepton plus four jets" event.

The top was found in 1995 by two different experiments, named CDF and D0, carried out by large collaborations at Fermilab. These were very difficult experiments for several reasons. There are countless events in a proton-antiproton collision. Most do not have the features required for a top event and are not even recorded on the computer. Less than one in a million events looks promising and is recorded. But only one in a few billion events turns out to be a top quark. How can we be sure that it is really a top and not something else? In fact, there are many so-called background events that could mimic the top. This means that all possible events must be well understood. To be absolutely certain, many top events must be collected before one can claim that the probability of a mere statistical fluctuation is very small. Understanding the background and having enough statistics are major concerns in particle physics experiments. In this respect, the "dirty" environment of proton-antiproton collisions is more demanding than the electron-positron collisions because there are six quarks involved, and these suffer strong interactions from a sea of virtual quarks and gluons.

9.3 THE CABIBBO–KOBAYASHI–MASKAWA MATRIX

The three families of left-handed $SU(2)_L$ doublets may be written as

$$\text{First family:} \quad \begin{pmatrix} \text{electron-neutrino} \\ \text{electron} \end{pmatrix}_L, \quad \begin{pmatrix} u \\ d_W \end{pmatrix}_L$$

$$\text{Second family:} \quad \begin{pmatrix} \text{muon-neutrino} \\ \text{muon} \end{pmatrix}_L, \quad \begin{pmatrix} c \\ s_W \end{pmatrix}_L$$

$$\text{Third family:} \quad \begin{pmatrix} \text{tau-neutrino} \\ \text{tau} \end{pmatrix}_L, \quad \begin{pmatrix} t \\ b_W \end{pmatrix}_L$$

The fields d_W, s_W, and b_W do not correspond to the d, s, and b quarks one measures, but rather to combinations of them:

$$d_W = V_{ud}\, d + V_{us}\, s + V_{ub}\, b$$

$$s_W = V_{cd}\, d + V_{cs}\, s + V_{cb}\, b$$

$$b_W = V_{td}\, d + V_{ts}\, s + V_{tb}\, b$$

Since the doublet structure controls the coupling with the $W^{\pm}$ gauge bosons, V_{ud} multiplied by g is the strength of the coupling of a W boson with one u quark and one d quark. Similarly, V_{us} determines how strong is the coupling of a W boson with one u quark and one s quark, and so on. This is shown in Figure 9.7. It is customary to combine all W quark couplings as in Table 9.1. Because it was due to the combined works of Cabibbo,

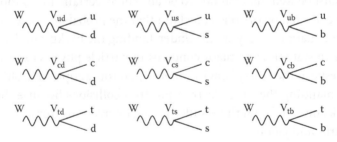

FIGURE 9.7 Representation of the couplings of the W gauge bosons with the quarks.

TABLE 9.1   Cabibbo–Kobayashi–
Maskawa (CKM) Matrix Determining the
Strength of the Interaction between the W
Gauge Bosons and the Quarks

W	d	s	b
u	V_{ud}	V_{us}	V_{ub}
c	V_{cd}	V_{cs}	V_{cb}
t	V_{td}	V_{ts}	V_{tb}

Kobayashi, and Maskawa, this is known as the Cabibbo–Kobayashi–Maskawa (CKM) matrix.

Each entry in Table 9.1 is a complex number, which can be viewed as a dial. As discussed in Appendix 3, each dial has a magnitude (size) and a phase (angle). Table 9.1 is a matrix in the technical sense. Because one must preserve probability, it is a unitary matrix. Unitary matrices are discussed in Appendix 4. As a result, of the nine magnitudes only three are independent. This occurs because the relations between d_W, s_W, and b_W (on the one hand) and d, s, and b (on the other) require the sum of the squares of the elements of any row in the Cabibbo–Kobayashi–Maskawa matrix to equal 1. Similarly, the sum of the squares of any column must equal 1. In fact, experiments have tried to test whether the sums of the squares really do equal 1 in order to see whether there might be some mixing with a fourth quark family. Within current experimental accuracy, there is no evidence for mixing with a fourth quark family.

The magnitudes of the elements of the matrix determine the rates of different decays. For example, the decay of the B⁻ meson (containing a b quark and a u antiquark) to the charm meson D⁰ (c quark and u antiquark) plus a negatively charged pion is proportional to $|V_{cb}|^2$, the square of the magnitude of the cb entry in Table 9.1. Similarly, the decay of B⁻ to a neutral pion and a negatively charged pion is proportional to $|V_{ub}|^2$. The dominant Feynman diagrams corresponding to these decays are shown in Figure 9.8. It turns out that V_{cb} is much larger than V_{ub}, so the decay rate to charm is much larger than the decay rate to final states without charm.

Once the decay rates for B mesons were measured it became apparent that V_{cb} was considerably smaller than V_{us}, just as V_{us} was smaller than V_{ud}. Similarly, V_{ub} was smaller than V_{cb}. Based on the decay rates, a rough estimate of the magnitudes of the matrix elements is shown in

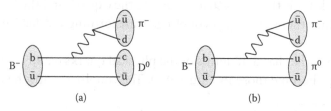

FIGURE 9.8 Dominant Feynman diagrams for the decay amplitudes of (a) $B^- \to D^0 \pi^-$ and (b) $B^- \to \pi^0 \pi^-$. The decay rates are proportional to the square of these amplitudes.

TABLE 9.2   Very Rough Picture of the
Hierarchy of the Cabibbo–Kobayashi–
Maskawa Matrix Elements

W	d	s	b
u	1	1/5	1/125
c	1/5	1	1/25
t	1/125	1/25	1

Note: Notice how the sizes diminish as one
goes away from the diagonal.

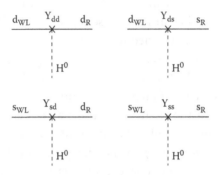

FIGURE 9.9   Representation of the couplings of the Higgs with the fields d_W and s_W.

Table 9.2. The three known families are just what one needs in order to accommodate the observed CP violation, as will be explained in Part B of this book.

9.4 YUKAWA COUPLINGS, MASSES, AND MIXING

We have mentioned in Chapter 8 that the Yukawa couplings determine how the quarks couple to the Higgs scalar. Some couplings are shown in Figure 9.9. To simplify, we consider only two quark families. As before, the quark d_{WL} is the quark that couples to the u quark and the W boson. The quark s_{WL} is the quark that couples to the c quark and the W boson.* Notice that there are Higgs couplings transforming a d field into an s field.

On the other hand, in terms of the physical quarks d and s, the Higgs couplings become those in Figure 9.10. The couplings with the physical

---

* Before we introduce the Higgs interaction the quarks are all massless. The quarks labeled d_{WL} and s_{WL} are distinguished by the way they interact. It is the Higgs vacuum expectation value that gives the quarks masses, and because the Higgs coupling has a piece that mixes d and s, the physical d and s quarks, the ones with definite masses, are mixtures of d_W and s_W.

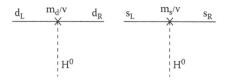

FIGURE 9.10 Representation of the couplings of the Higgs with the physical quarks d and s.

quarks only couple a d quark with itself, or an s quark with itself. This is as required by the Higgs mechanism, which gives masses to the quarks. Since the mass is determined by the Higgs coupling times the Higgs vacuum expectation value, v, one can write the coupling as m/v. Thus, the Cabibbo angle and the mixing in Figure 9.7 originate in the Yukawa couplings.

Similarly, the Cabibbo–Kobayashi–Maskawa matrix of the standard model with three quark families has its origin in the Yukawa couplings. This stresses the theoretical importance of Yukawa couplings: they determine the fermion masses, their mixing, and also CP violation.

NOTES

1. In formulas, $g_{ud} = g_\mu \cos\theta_C$ and $g_{us} = g_\mu \sin\theta_C$.
2. We use $d_W = d \cos\theta_C + s \sin\theta_C$.
3. We use $s_W = -d \sin\theta_C + s \cos\theta_C$.
4. The diagram in Figure 9.3a is proportional to $(g^2 \cos\theta_C \sin\theta_C)$, while the diagram in Figure 9.3b is proportional to $-(g^2 \cos\theta_C \sin\theta_C)$. If m_u were equal to m_c, the minus sign difference would mean that one diagram would cancel completely the contribution from the other.
5. We have chosen to simplify things and write the charge +2/3 quarks already in their mass basis. This forces an apparent lack of symmetry between the charge +2/3 and charge –1/3 quarks. In actual fact, the theory is initially written in a weak basis for both charge +2/3 and charge –1/3 quarks. The Yukawa couplings mix all quarks with the same charge. One rotates u_W and c_W into the mass basis through an angle θ_u; one rotates d_W and s_W into the mass basis through an angle θ_d; the Cabibbo angle is given by $\theta_C = \theta_d - \theta_u$. Said otherwise, the Cabibbo angle arises because of a clash between the following two requirements: (1) one must take the quarks into their mass basis, which in general requires a different angle for the left-handed charge +2/3 and charge –1/3 quarks, and (2) these left-handed quarks are connected by the weak interactions, and thus should rotate in the same way. In our presentation we have followed the established convention of rotating the charge +2/3 quarks into their mass basis from the start. Then the mismatch is blamed on the charge –1/3 quarks, through the Cabibbo angle θ_C.

CHAPTER 10

Our Current View of Nature's Building Blocks

(What We Have Learned So Far)

W E HAVE DISCUSSED THE historical roots of our current knowledge about particle physics. This is a good place to contemplate the present picture before we discuss some of the important issues being explored today in the following parts of the book.

10.1 THE FOUR INTERACTIONS OF NATURE

Physics is the science that studies the universe at all scales. It covers phenomena ranging from reactions with pointlike leptons and quarks to reactions with nuclei and atoms; from interactions between atoms in small molecules to protein folding and the interaction of four macromolecules to form DNA (the very source of life); from the transport of nutrients across cell membranes to the transport of information in our body; from climate changes in our planet to its motion across our solar system; from events in our Milky Way to the formation of clusters of galaxies and the fate of the universe as a whole. Physics covers all these subjects. If you like any of these phenomena, you like physics.

There are millions of phenomena that prod the curious mind. Perhaps the most remarkable achievement of particle physics is this: we now know that at the root of all these countless phenomena there are only four interactions. And, at a fundamental level, we can describe these interactions mathematically. Behind the confusion of everyday life, there are four simple rules. This is shown schematically in Figure 10.1. At a given scale (length),

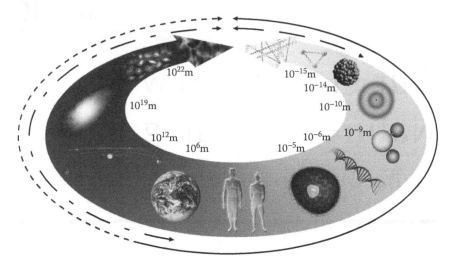

FIGURE 10.1 The four interactions of nature. The scales at which they become relevant appear as a dash-dotted line (gravity), a solid line continued by a dotted line (electromagnetism), and a dashed line (strong and weak interactions). (From R.S. Mackintosh, J.S. Al-Khalili, B. Johnson, and T. Pefia. Canopus, Nucleus: A trip into the heart of matter. Canopus Publishing, Jamie Symonds. With premission.)

there may be some interactions that are less relevant than others. Different interactions have a bigger impact on some length scales than on others. For example, gravity is especially relevant for all scales equal to a planet's size or larger. But gravity is minutely small, and thus completely irrelevant, when one studies individual atoms. Conversely, the electromagnetic interaction is relevant for all phenomena at planetary and smaller scales. It is relevant to our cell phone, for the transmission of information in our brain, and it must also be considered when studying reactions involving individual leptons and quarks. The electromagnetic interaction turns out to be important also to our knowledge of phenomena at larger scales, not because it directly influences large-scale phenomena, but because almost all information we get from other portions of the universe comes from the electromagnetic radiation they emit.*

Electromagnetism is behind the interaction between water molecules; it explains why the negatively charged electrons orbit the positively charged nucleus, and it explains how electromagnetism leads one oxygen atom and

---

* The exceptions are neutrinos and cosmic rays. The beginning of neutrino astronomy is discussed in Part C. Cosmic rays consist primarily of protons with some heavier nuclei mixed in and may come from distant galaxies where they have been accelerated to high energies.

two hydrogen atoms to come together, forming the water molecule. But electromagnetism leaves us with a puzzle once we look inside the nucleus. The nucleus of oxygen, for example, has eight protons inside a very small volume. These protons, all with the same electric charge, experience an enormous electric repulsion that should destroy the oxygen nucleus. This is not observed. So, we require an interaction that is stronger than the electric repulsion and which joins the protons together. Because the interaction has to be stronger than the electromagnetic repulsion, physicists named it the strong force.

Natural oxygen has eight protons and several possible numbers of neutrons: eight, nine, and ten. Their abundances are 99.757, 0.038, and 0.205%, respectively. Clearly, only a few oxygen atoms in each ten thousand have nine neutrons. But, because there are millions and millions of oxygen atoms in our body, there are still many millions of this type. Looking at these neutron numbers, it is natural to wonder whether we could produce an oxygen nucleus with eleven neutrons (in addition to the eight protons all oxygen nuclei have). This is indeed possible, but such a nucleus only survives for about 27 seconds. After this time it disappears completely, and in its place appear one electron, one antineutrino, and one nucleus of fluorine having nine protons and ten neutrons. In this process, one neutron from the oxygen nucleus decayed into one proton, one electron, and one antineutrino. Such decays cannot be explained by the gravitational, electromagnetic, or strong interaction. Because this requires an interaction weaker than the electromagnetism, physicists dubbed it the weak interaction.

The strong and weak interactions are relevant for subatomic processes. They are relevant at scales smaller than the atom. If the energy of a particle is very large, it probes very small distances. In stars and the very early universe, the temperature is very high and particles have very large energies; for these phenomena the weak and strong interactions are also relevant.

A word of caution: Since all matter is made of the same basic constituents, since there are only four interactions, and since we have a solid mathematical framework encompassing all fundamental particles and their interactions, one might be tempted to believe that all phenomena can be calculated. This is wrong! For example, we know that the interaction between water molecules is electromagnetic in nature. But in any small amount of water there are millions and millions of molecules, making it very difficult to perform reliable calculations. The best one can do is model the interactions within the liquid. But then the results depend in a very sensitive way on the assumptions that went into building the model. There are some properties that we can calculate from fundamental theories, but we cannot hope to calculate most properties of systems with many particles from such first principles.

10.2 THE FUNDAMENTAL BUILDING BLOCKS OF MATTER

Throughout recorded history humans faced the same tantalizing questions. What are we made of? Do we share components with other living creatures? Do we share components with a rock? And with the stars? Is there some simplicity behind this apparent complexity? How many building blocks are there?

We now know the answers to many of these questions. At a very fundamental level, all things have the same components. Almost all that we see is made up of indivisible electrons, up quarks, and down quarks. These are shown in the first line of Table 10.1, together with the electron neutrino. The fact that Table 10.1 has further lines (so-called families or generations) led to I. I. Rabi's famous quote: "Who ordered that?" Due to Einstein's $E = mc^2$, these heavier fundamental particles are only produced through sources of considerable energy, such as cosmic events or deliberately man-made laboratories. We do not know why there should only be three families. This is a question for the twenty-first century; it is a question for the young reader.

For every particle in Table 10.1 there is an antiparticle with opposite charge. As we saw, the antielectron (positron) was first invented by Dirac for consistency of a mathematical equation, and only later discovered experimentally. This might also be a message for the young reader: a mathematical equation is merely a structured way of thinking about real things. A structured procedure so reliable that you can trust it even when

TABLE 10.1   Fundamental Matter Particles

| Family | Leptons | | Quarks | |
	Charge –1	Charge 0	Charge –1/3	Charge +2/3
1st	e^- Electron	ν_e Electron neutrino	d Down quark	u Up quark
2nd	μ^- Muon	ν_μ Muonic neutrino	s Strange quark	c Charm quark
3rd	τ^- Tau	ν_τ Tau neutrino	b Bottom quark (or beauty quark)	t Top quark (or truth quark)

it implies the prediction of a previously unseen particle. It pays to learn mathematics; the physics Nobel Prize that Dirac won for his prediction is currently worth 10 million Swedish kronor (currently around US$1.6 million or €1.1 million).

We are made of particles, not antiparticles. In fact, the visible universe is made up almost exclusively of particles. Why is the excess of particles over antiparticles so overwhelming when they seem to be equally relevant in the theory? There are many ideas to solve this puzzle, but so far, there is no consensual solution. This is another question for the twenty-first century.

Leptons do not feel the strong interaction. The electron, muon, and tau have a property known as electric charge that makes them feel the electromagnetic interaction. There are two types of electric charges: positive and negative. Antiparticles have the same mass as the corresponding particle. They also have the same absolute charge, but with opposite signs. Conventionally, the electron has a negative electric charge and the positron (the antielectron) has a positive electric charge. The neutrinos have no electric charge, and thus cannot feel the electromagnetic force. Neutrinos only feel the weak force, making them extremely difficult to detect.

Quarks have electric charge, and thus feel the electromagnetic interactions. But quarks also have a property that makes them feel the strong interaction, the so-called strong charge. Because there are three types of strong charges, in analogy with the three primary colors, the strong charges are known as "color" and assigned the tags red, yellow, and green. The antiquarks carry the colors antired, antiyellow, and antigreen. Since the strong interaction keeps increasing with distance, quarks are never found as free particles. They always appear either in a quark-antiquark combination (known as a meson), in a combination of three quarks (known as a baryon), or in a combination of three antiquarks (known as an antibaryon). Mesons, baryons, and their antiparticles are known collectively as hadrons. The color assignments of the quarks inside hadrons are such that the overall strong charge is zero; we say they are (color) white.

Two further notes. The different types of quark (down, up, strange, charm, bottom, and top) are known as the quark flavors. Leptons and quarks are known collectively as the particles of matter. This will distinguish them from the particles that mediate interactions, to be discussed below.

10.3 INTERACTIONS ARE MEDIATED BY PARTICLES

Our current theories involve quantum fields, which display both the particle and wave character of what we call elementary particles. In addition to those listed in Table 10.1, there are the W and Z bosons and the photon as well as the gluons. These play the special role of mediating the weak, electromagnetic, and strong interactions.

Their mass and coupling strength will determine how these interactions are perceived. When we consider some interactions, we are interested in a few questions: How strong is that interaction compared to others? How far away can this interaction be felt? Does having more matter mean a stronger interaction through a cumulative effect? Admittedly, these are qualitative notions. We have not defined them in any precise fashion. But we may get some qualitative answers by looking at the messenger particles.

The particle associated with the electromagnetic interaction is the photon, and it has zero mass. As a result, it travels at the velocity of light in all frames of reference. In Chapter 4, we noted that the photon coming off the vertex of a free electron must be viewed as a virtual particle. It cannot be produced in a way consistent with energy and momentum conservation and consistent with its existence as a real particle respecting the energy-momentum-mass relation of special relativity. Using the uncertainty principle, we were able to explain why the electromagnetic interactions have infinite range.*

The strength of the electromagnetic interaction at a vertex is given by the charge of the particle. All fundamental particles have charges related to the electron's charge, which therefore sets the overall strength of electromagnetic interactions. If you accumulate many charges, you increase the strength of the interaction. This is roughly what happens to the clouds during a thunderstorm. The result is a powerful lighting discharge. However, under normal circumstances, objects have as many negative as positive charges. They are electrically neutral. There may be only some residual electric interaction due to the uneven distribution of charge within the object. For example, the electric interaction between two water molecules is due to the uneven charge distribution; there is a slight excess of negative charge close to the oxygen atom and a slight excess of positive charge close to the two hydrogen atoms within each molecule. But this makes the inter-

---

* In practice, infinite range means that the force falls off as one over the square of the distance, and finite range means exponential fall-off.

action smaller than that between two electrons. The electric interaction between large objects is typically even smaller.

In analogy to the photon, there are three messenger particles associated with the weak interactions. Two form a particle-antiparticle pair. Their mass is roughly eighty-five times larger than the mass of the proton, and they have opposite electric charges, equal in magnitude to the charge of the electron. They are known as W bosons and represented by W^+ and W^-. Because they are charged, they interact with the photons. This interaction may be described with the help of the Feynman diagram in Figure 10.2a. There is a third carrier of the weak interaction, known as the Z boson, represented by Z, and whose mass is roughly 13% larger than the mass of the W bosons. It is a neutral particle (its electric charge is zero), and thus it does not interact with the photon. It does, however, interact with the other weak gauge bosons in a variety of ways. One is shown in Figure 10.2b.

The strength of the weak interaction at a vertex is given by the weak charge, which is somewhat larger than the electric charge. However, because the virtual W or Z that has to mediate the interaction between two vertices is so massive, the actual interaction strength at low energies is much smaller than the electromagnetic. At energies well above $M_Z c^2$, the weak and electromagnetic interactions are comparable in strength.

The carriers of the strong force are a set of eight particles known as gluons, because they glue together the quarks within hadrons. They are massless, like the photon. Unlike the photon, which does not have electric charge, the gluons do carry strong charge in a special way. For instance, we may have a gluon with red/antigreen attributes. Upon reaching a green quark, the gluon will be absorbed, transforming the green quark into a red quark. Because the strong charge is known as color, the theory of the strong interactions was named quantum chromodynamics (QCD). Since gluons are massless, we might expect the range of the strong force to be infinite, just as for the photons. However, gluons are subject to the strong interaction, so they too cannot be found as free particles. As one tries to

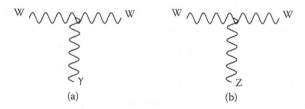

FIGURE 10.2   Feynman diagrams for some interactions among gauge bosons.

pull a single quark or a single gluon out, the energy in the gluonic field becomes so huge that eventually a particle-antiparticle pair will appear and new hadrons will form instead. A well-known analogy resorts to elastic strings. The strong force is like a spring. When you start to pull, the elastic string makes a small force on your hand. But as you move farther and farther away, the string pull gets stronger and stronger. Eventually, you are making such force that the string breaks into two smaller strings.

This is related to the issue of running coupling constants. At smaller scales (higher energies), the strong coupling constant is small and the quarks are essentially free. This is known as asymptotic freedom and was explained by Gross, Politzer, and Wilczek in 1973. Conversely, at larger distances (smaller energies) the strong coupling constant becomes very large, thus confining the quarks and gluons within hadrons. This is known as confinement and has severe consequences. We recall that calculations are made in quantum field theory through an approximate method. We start with free fields and we consider how the interactions change the result. First, we consider the effect of a single vertex, then two, then three, and so on. Since each vertex involves the coupling constant, we get a series of terms, each proportional to successive powers of the coupling constant. For example, the tree-level diagrams we saw when discussing QED are proportional to the second power of the coupling constant. If the coupling constant is small, each new term in the series is smaller than the preceding term (when renormalization is taken into account). But, if the coupling constant is very large, the new terms become larger than the ones we have calculated before. Our calculations become useless because the terms we have yet to calculate represent a much larger proportion of the final result. We say that the perturbative expansion breaks down. This is precisely what happens with the quarks inside hadrons. Because we know how to write the theory of quark-gluon interaction, you might expect that we would be able to calculate everything related to the way in which quarks bind within hadrons. We cannot!

Theoretical calculations of processes involving hadrons are actually carried out with quarks (and possibly also gluons) in the final states. The way quarks bind into hadrons is known as hadronization. As noted in Chapter 6, for very high energies it is possible to detect narrow jets containing many particles (like pions), which can be understood as originating from a quark or a gluon produced in the original interaction. However, in processes at lower energies, quark-level calculations will not do; we must go from our quark calculation to the observed hadrons. One way is to try to use QCD to go from quarks to hadrons. It requires extremely elaborate computer calculations, which are still of very limited accuracy in many cases. This can lead to difficulties in trying to accurately test our theories of the weak and strong interactions of hadrons. Nevertheless, by comparing and combining the results from many experiments, we have confirmed to quite good accuracy our present standard model.

The situation with gravity is very different. In 1912 Einstein proposed a theory of general relativity, encompassing Newton's law of gravitation as a particular case. This changed completely our view of space-time as a static theater where all particles and interactions perform; space-time itself became a central actor in the plot. We suspect that this theory is also quantized, and there will be an associated massless quantum, known as the graviton. Because gravity has an infinite range, the graviton must be massless. Unfortunately, no one has been able to produce a predictive theory of quantum gravity. In 1984 Green and Schwartz showed that a theory known as superstrings was a candidate for quantum gravity. Superstrings have remained a candidate for over 20 years, and despite great progress, there is no experimental proof that it is the right idea.

Gravity is an incredibly feeble interaction at the scale of neutrons and protons. Thus, it is not included in the so-called standard model of particle physics, which concerns only the electromagnetic, weak, and strong interactions. But, because it is always attractive, gravity is cumulative. Gravity is proportional to an object's mass. A rock has an enormous number of fundamental particles. Each has a tiny mass and a small electric charge. Because there are as many positive as negative charges in a rock, making it electrically neutral, it has no electric interaction with another rock. The gravitational interaction between two rocks is also small, but for a different reason. The masses of individual particles are so tiny that, despite the enormous number of them within a rock, the overall mass is still not large enough for the cumulative effect of gravity to become noticeable. However, because it is cumulative, when rocks are accrued into planets, gravity has a measurable effect.

A summary of the properties of messenger particles is shown in Table 10.2. This table is adapted from the "Table of Particle Properties," edited by the Contemporary Physics Education Project.[1] The "Table of Particle Properties" can be found in every corner of the world and has been responsible for the decision of many young students to embrace a scientific life. The draft for that table was made by Helen Quinn on her kitchen floor while juggling other difficult problems. Helen Quinn is a female scientist

TABLE 10.2   Messenger Particles

Interaction	Messenger Particles	Strength Relative to Electromagnetic, for Two Up Quarks at	
		10^{-18} m	3×10^{-17} m
Weak	W^+, W^-, Z	0.8	10^{-4}
Electromagnetism	Photon	1	1
Strong	8 gluons	25	60
Gravity	Graviton	10^{-41}	10^{-41}

Note: Electromagnetism and gravity have an infinite range because the photon and graviton are massless. Gluons are also massless, but the range of the strong interaction is curtailed by confinement. At a large distance, the other interaction is weak because its W and Z boson carriers are massive. There exists no direct evidence for the graviton.

who has made important contributions to physics, physics education, and physics administration (she is a former president of the American Physical Society), despite the fact that during much of her career physics research was still a mostly male-dominated world.

10.4 THE STANDARD MODEL AND LARGE LABORATORIES

There is a theory describing all the matter particles summarized in the previous sections, along with their weak, electromagnetic, and strong interactions. These are described in the mathematical framework of a relativistic quantum field theory. For three decades, this theory has been subject to ever more precise tests. Its predictive power is so uncanny that it is known as the standard model of particle physics (SM, for the specialists).

In this theory, the weak and electromagnetic interactions are unified. In the 1850s Maxwell finished the job started by Faraday of unifying the electric and magnetic interactions (and optics) into the electromagnetic interaction. In the 1960s Glashow, Weinberg, and Salam showed that the electromagnetic and weak interactions could be considered two aspects of the same theory, known as electroweak, for which they were awarded the Nobel Prize in 1979. This theory only aroused great interest in 1972, after 't Hooft and Veltman showed that the theory is renormalizable, in the same sense as QED, discussed before. 't Hooft and Veltman won the Nobel Prize in 1999.

This second unification was no trivial undertaking, given the differences of range and strength of both interactions. As we pointed out, part of the reason has to do with the large masses of the W and Z bosons. The corresponding theory implied the existence of these so-called gauge bosons, which were discovered by two experiments at CERN in 1983. As with Dirac's positron, belief in a mathematical description led again to the prediction of particles that truly do exist.

The great difference between the weak and electromagnetic interactions arises from the fact that the W and Z bosons have large masses, whereas the photon is massless. In the standard model the masses arise as a result of the interaction with the Higgs field. The theory requires that in high-energy collisions Higgs bosons, the particles associated with the Higgs fields, should be produced. So far no Higgs bosons have been observed; this may be considered the missing link of the standard model. If the theory is correct, Higgs bosons will be discovered at the Large Hadron Collider (LHC), the new accelerator at CERN in Geneva. In this case, it will have taken more than four decades to uncover all the particles of the standard model. It will be even more exciting if the Higgs boson is not found, since that will mean there is still some fundamental new physics to be discovered.

Testing theories of particle physics and making new discoveries today involves large machines that accelerate particles to very high energies. When the first new particles, the positron, muon, and pion, were discovered, the experiments depended on cosmic rays. In order to do controlled experiments, the first accelerator was made in England by Cockcroft and Walton (Nobel Prize, 1951) in 1932. This involved accelerating protons across an electric field. To get to larger energies, E. O. Lawrence (Nobel Prize, 1939) invented the cyclotron, in which protons were sent around and around in a magnetic field with an electric field giving them a push to higher energy after each turn. In order to take account of special relativity, the electric pulses had to be carefully timed leading to the synchrocyclotron. Over the years, larger and larger accelerators were built, culminating in the Tevatron at Fermilab with a ring of 6.3 km in which protons and antiprotons were collided together, each with an energy of 1 Tev. It was there that the last of the elementary particles was discovered, the top quark. Now an accelerator with seven times the energy, the LHC, is entering into operation at CERN. Important discoveries have also been made at accelerators that collide electrons with positrons. While these do not involve energies that are as high, the experiments are often easier to analyze. Discussed before were the discoveries of charm and the tau-lepton at the SLAC at Stanford. More recently, the BABAR experiment at SLAC and the BELLE experiment at KEK in Japan have made definitive studies of B mesons. The LEP experiment at CERN accelerated electrons and positrons, each to an energy up to 100 Gev, providing precision tests of the standard model and a lower limit on the Higgs mass.

CERN (originally the acronym for Conseil Européen pour la Recherche Nucléaire) is an enormous scientific European facility situated on the border of France and Switzerland. It has around 2,500 staff, but it is visited by around 8,000 scientists from all over the world (roughly half the world's particle physicists). It is interesting to look at the breakdown of CERN's staff. Contrary to what one might expect, less than one-fifth are scientists. About as many are support staff involved in cleaning, catering, secretarial duties, etc. The majority are engineers or technicians. And there are even a few people whose skills are so precious that one might call them artisans. There are several important lessons here. The media usually portray science as the business of some nutty and isolated professor to whom they attribute center stage. This is not at all how science is made. Science is made by all sorts of people, with all sorts of personalities and skills. It is when you join such disparate talents that humankind transcends itself. Another lesson is especially important for young people: if you do like science and are willing to put in the long hours required, you are welcomed. Maybe you are especially adept at mathematics; you are welcomed to science. Maybe you do not really love math, but you have great practical skills; you are welcomed to science. Whatever your skills, you may be useful to science.

LHC is a 27 km ring in which protons are accelerated very close to the speed of light and then collided. Due to Einstein's $E = mc^2$, the enormous energy resulting from this collision will show up in the form of myriad particles. The first aim of such collisions is to produce the Higgs particle.

To observe these particles, the enormous ATLAS and CMS detectors have been built. In essence, these are cameras that photograph particles. But these cameras are six stories high and weigh around ten thousand tons. They are technological powerhouses. The accelerator itself is also a technological marvel. For example, the magnets involved in guiding the beam of protons are gigantic and placed at –271.3°C, just 1.9° above absolute zero. LHC has truly the largest fridge in the world. The operation is so complex that it involves close collaboration between a thousand scientists from all over the world.

You can immediately guess one problem with this collaboration. How can scientists from Taipei, Sydney, London, Lisbon, San Francisco, Berlin, Pittsburgh, and a hundred other cities manage together a six-story building placed on the frontier between Switzerland and France? It was precisely to solve this problem that the World Wide Web was invented by Tim Berners-Lee at CERN in 1990. Yes! The Web you use to buy books, concert tickets, and groceries was invented by particle physicists to solve a practical problem. It was not invented by crazy scientists secluded on their ivory towers. It is when you congregate many skills into one project that truly remarkable advances are possible.

A lot has been said about the cost of these facilities. We would like to put forth the following argument. The Web is free because it was invented in such an international nonprofit organization. Imagine that it was instead patented by a private company. Each time you clicked on a link you would pay some amount to that company. This would certainly curtail your fun and probably the world economy. Similar success stories exist at SLAC (the Stanford Linear Accelerator Center, which housed the first Web server in the United States) in California; Fermilab (Fermi National Accelerator Laboratory) near Chicago; KEK (High-Energy Accelerator Research Organization) in Tsukuba City, Japan; and the other major facilities. Because they push the frontier of science, they pay themselves in technological transfers. But do not delude yourself. Outstanding as these side benefits are, the most important revenue is still the science being produced—the satiation of the thirst for knowledge that defines our human condition.

NOTE

1. More information may be found at the Web site http://www.cpepweb.org/.

PART B

CP Violation: The Asymmetry between Matter and Antimatter

Quantum electrodynamics based on Dirac's theory of the electron has a symmetry between particles and antiparticles, between electrons and positrons, called charge conjugation, C. There was an open question whether something similar held for the proton. However, it was only in 1955 that the antiproton was first detected, and only years later was it possible to create antihydrogen atoms by combining antiprotons with positrons. The C symmetry also seemed to hold for protons and antiprotons. Thus, the C symmetry came to be called the symmetry between matter and antimatter.

When the violation of parity was discovered, as discussed in Chapter 7, it was found that C was also violated, but that CP invariance appeared to hold. Thus, CP symmetry could be considered the symmetry between matter and antimatter. If this symmetry really existed, a question arose as to why the universe appeared to be made of matter without antimatter. One possibility suggested was that the Milky Way and nearby galaxies resulted from a slight fluctuation yielding a net matter content, and that other galaxies contained antimatter. However, it was known that there is material between galaxies. In this intergalactic region there would be matter annihilation with antimatter, which could be detected from the gamma rays they produced. No such evidence was found. The visible universe appears to be entirely made of matter.

In 1964 the small violation of CP invariance was discovered. This led Andrei Sakharov to suggest that the universe might have started out with symmetry between matter and antimatter and then, as a result of CP violation, some processes occurred in the early universe that left a net amount of matter. This process is known as baryogenesis. There is a quantity known as baryon number, B, where protons and neutrons have B = 1 and their antiparticles have B = −1. In order for an asymmetry between matter and antimatter to appear, Sakharov had to assume that there were processes that violated the baryon number. Finally, if there were thermal equilibrium, there would be as many events producing matter from antimatter as there would be in the reverse direction. Any asymmetry would be washed out. Therefore, lack of thermal equilibrium is also required.

The standard model has all these ingredients. It has CP violation, it has processes that violate the baryon number,* and it has a phase transition at the time the Higgs field develops a nonzero vacuum expectation value, which may lead to the required lack of thermal equilibrium. However, detailed calculations indicate that the CP violation in the standard model is not sufficient to produce the observed baryon asymmetry.

There are many theoretical ideas building upon this suggestion of Sakharov, and they all depend on CP violation. This is one of the reasons for the interest in trying to understand the origin of CP violation.

---

* These are nonperturbative effects, meaning that they cannot be calculated within the perturbation technique described in Chapter 4. But they only become significant at very high energies, and they are of no importance at experimentally accessible energies.

CHAPTER 11

CP Violation in Kaon Decays

T HE VIOLATION OF CP symmetry was first observed in 1964 in a study of the decays of the neutral kaon mesons, as noted in Chapter 7. The K0 meson may be considered the combination of a d quark and an s antiquark. Applying the CP transformation, we obtain a d antiquark and an s quark, which is known as the meson K0bar (pronounced "kay-zero-bar").

Therefore, neither K0 nor K0bar has a definite CP quantum number. But one can define quantum mechanical wave functions that describe a superposition of K0 with K0bar. In the designation of quantum mechanics, we call these wave functions states. For example, we can construct the superposition

$$K_+ = \frac{1}{\sqrt{2}}\left(K^0 + \bar{K}^0\right),$$

$$K_- = \frac{1}{\sqrt{2}}\left(K^0 - \bar{K}^0\right).$$

We stress that the mere possibility of defining these states hinges on the quantum mechanical opportunity for superposition. These states have a probability of 50% of being detected as composed of an s quark and a d antiquark, and a 50% chance of being detected as composed of an s antiquark and a d quark. Because CP interchanges K0 and K0bar, it is easy to see that

$$K_+ \xrightarrow{\;CP\;} + K_+,$$

$$K_- \xrightarrow{\;CP\;} - K_-$$

This means that K_+ has the quantum number $CP = +1$, while K_- has the quantum number $CP = -1$. If CP were conserved, these would be the states with definite masses and lifetimes.

Neutral kaons can decay into two and three pions. Let us consider first the decays into two pions. Because the kaons and pions do not have spin, conservation of angular momentum forces the relative orbital angular momentum of the two pions originated in a kaon decay to be zero. Combining the C and P properties of this system, we conclude that the final states $(\pi^+ \pi^-)$ and $(\pi^0 \pi^0)$ have $CP = +1$. Thus, if CP were conserved, K_+ could decay into $(\pi^+ \pi^-)$ or $(\pi^0 \pi^0)$, while those decays would be forbidden to K_-. Next, we consider the decays into three pions: $(\pi^+ \pi^- \pi^0)$ and $(\pi^0 \pi^0 \pi^0)$. Because the kaon has spin zero, we expect the three pions to emerge with no relative angular momentum and the three pion state has $CP = -1$.* Thus, if CP were conserved, K_- would decay into three pions, while this decay would be forbidden to K_+.

Now the mass difference between the neutral kaon and two pions is much larger than the mass difference between the neutral kaon and three pions. As a result, it is much easier for the neutral kaons to decay into two pions. All other things being equal, we conclude that K_+, which can decay into two pions, should disappear much faster than K_-, which cannot. We predict that there will be two neutral kaons with very different lifetimes. This is indeed what one finds experimentally. There are two neutral kaons. One is short lived (denoted by K_S), with a lifetime of the order 9×10^{-11} seconds, while the other is longer lived (denoted by K_L), with a lifetime of the order 5×10^{-8} seconds. If CP were conserved, K_S would coincide with K_+ and be allowed to decay into two pions, while K_L would coincide with K_- and would never decay into two pions.

The fact that K_L lives much longer than K_S has a very interesting consequence. Imagine a beam that has initially equal amounts of K_S and K_L. At this stage some of the K_S will decay into two pions. As the beam travels, the K_S component will disappear very fast. After some distance we are left with the K_L component, which lives longer. And, if CP were conserved, no

---

* The minus sign arises because each of the pions has intrinsic parity equal to –1, and –1 cubed is –1.

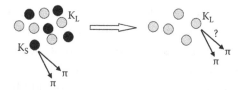

FIGURE 11.1   An experimental search for CP violation. A beam of K_S (in black) and K_L (in grey) travels along. After some distance, essentially all of the K_S component will have disappeared, some of it into two pions. If the surviving K_L also decays into two pions, then CP is violated.

K_L would decay into two pions. Conversely, if the surviving K_L particles decay into two pions, we have a clear sign of CP violation. This experiment is shown schematically in Figure 11.1. The results of this experiment were announced in 1964 by Christenson, Cronin, Fitch, and Turlay. While the K_L decayed primarily into three pions, occasionally they observed the decay into two pions. Because K_L decays into both two and three pions, which are final states of opposite CP, CP violation had been clearly identified. The decay of K_L into ($\pi^+ \pi^-$) is one hundred thousand times less likely than the decay of K_S into the same final state. Thus, in the kaon system, CP violation is a small effect. In technical articles, CP violation in this experiment is measured by a parameter $|\varepsilon|$ of order 0.002.* Cronin and Fitch were awarded the Nobel Prize for this discovery in 1980.

One further example might be helpful. Because there is no other particle of equal mass, if CP were conserved, K_L would transform into itself under CP with a minus sign. It can decay into $\pi^+ e^- \bar{\nu}_e$ and into the C-conjugated final state $\pi^- e^+ \nu_e$. When considering the decay rates into these final states, we sum over all possible momenta and spin configurations. This means that the decay rates are related by CP.† Said otherwise, a difference between the decay rates of K_L into $\pi^+ e^- \bar{\nu}_e$ and into $\pi^- e^+ \nu_e$ is a sign of CP violation. Experimentally, π^- appears more often than π^+. This signal of CP violation is found to be very small and can be directly related to kaon mixing and correctly calculated in terms of the parameter $|\varepsilon|$.

The question then was to determine the origin of this CP violation. One possibility suggested by one of us (Lincoln Wolfenstein) was that this might be due to a new interaction much weaker than the normal weak

---

* Assuming the validity of the CPT theorem discussed in Chapter 7, this is also a sign of a small violation of time reversal.

† The minus sign coming from the CP transformation of K_L is irrelevant for the decay rates because it gets squared in the calculation.

interaction; this came to be called the superweak theory. It was perhaps because of this possibility that the Weinberg theory of the weak interaction with only four quarks (see Chapter 8) could be accepted as a good theory, even though it could not explain CP violation.

According to the superweak idea, the CP violation has to do with the mixing of K0 and K0bar, so that the particle with definite mass and lifetime K_L was not purely K_- (CP odd), but had a small admixture of K_+ (CP even), which allowed the small decay of K_L into $\pi^+ \pi^-$.

Since K0 has strangeness S = +1 and K0bar has S = −1, it is necessary to change S by two units to mix them. The standard model weak interaction changes S by one unit so that it is necessary to apply the weak interaction twice to do the mixing. This is illustrated by the diagram in Figure 11.2. In the original standard model with four quarks the quark labeled q is either the charm quark c or the up quark u. The result is a very small mass difference because the weak interaction is applied twice, but there was no CP violation.

The superweak proposal was to add a new interaction that directly violated S by two units, and yet its effect was smaller than the second-order weak interaction of the standard model, as shown in Figure 11.2. A prediction of the superweak theory was that, except for the small CP violation that had been observed in the K0 system, there would be no other signal of this new interaction, no other CP violation. That prediction seemed true for over 25 years after 1964.

The key was to discover CP violation that could not be blamed entirely on the mixing but had to do with the decay process itself; this is called direct CP violation. The original CP violation experiments observed the K_L decay into $\pi^+ \pi^-$. Now the goal was to look at the more difficult decay to $\pi^0 \pi^0$, which required detecting and accurately measuring the four photons that came from the decays of the two neutral pions. One could then

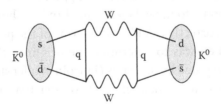

FIGURE 11.2   Box diagram that leads to K0-K0bar mixing in the standard model. In the original form of the standard model, q was either u or c, and there was no CP violation. In order to explain CP violation, it is necessary to add the diagram with t as q.

compare the ratio of those two decay modes in the K_L decay to the same ratio for K_S decay. The two ratios should be equal if the decays are both K_+ decays. The difference is measured by a new parameter, ε' (pronounced "epsilon-prime").[1] The parameter epsilon-prime is now a measure of direct CP violation.

Large-scale experiments to measure epsilon-prime were carried out at Fermilab and CERN in the 1980s, but the results had large errors and didn't agree. This led to new experiments: KTev at Fermilab and NA48 at CERN. One essential ingredient is that the K_L decays and K_S decays have to be observed in the same detector. There had to be very large and accurate electromagnetic calorimeters to measure the energies of the four photons: the KTev experiment used cesium iodide crystals, while NA48 used liquid krypton. It was only in the late 1990s that the two experiments began to give consistent results and provided a clear nonzero value for epsilon-prime, which was equal to 0.002 times the value of epsilon. While this value was very small, it left no doubt that there was a direct CP violation.

NOTE

1. Mathematically one wrote $A(K_L \to \pi^+ \pi^-)/A(K_S \to \pi^+ \pi^-) = \varepsilon + \varepsilon'$, $A(K_L \to \pi^0 \pi^0)/A(K_S \to \pi^0 \pi^0) = \varepsilon - 2\varepsilon'$.

CHAPTER 12

The Cabibbo–
Kobayashi–
Maskawa Matrix

CP Violation in the Standard Model

I N 1973, EVEN BEFORE the discovery of the b and t quarks, it was proposed in one paragraph of a paper in the *Progress of Theoretical Physics* that if there were six quarks instead of just four, it would be possible to extend the Weinberg model to include CP violation. The authors were Kobayashi and Maskawa. This has become the standard model of CP violation, and in 2008 Kobayashi and Maskawa were awarded the Nobel Prize in physics.

The starting point is the fact that any field theory of spin 1/2 particles also includes antiparticles, as originally discovered by Dirac. The other crucial point is that the amplitude of any process like a weak decay is described by a complex number. As discussed in Appendix 3, a complex number may be defined by the length of an arrow and a dial that gives the direction, like the hand of a clock. The direction is called the phase of the complex number. Phases play an essential role in the theory of CP violation.

The fundamental theory of weak interactions contains two terms, as illustrated in Figure 12.1.* The term shown in Figure 12.1a is the one responsible for the beta decay of the neutron, taking d_L into u_L and W^-. It is proportional to V_{ud}. V_{ud} is a complex number; it can be viewed as an

---

* Recall that a bar over a quark letter designates the corresponding antiquark. Thus, $\bar{u}$ is the u antiquark, for which we will sometimes also use the notation ubar.

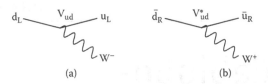

FIGURE 12.1 Two terms arising from the standard model description of weak interactions: (a) left-handed d quark going into left-handed u quark and W^-; (b) right-handed d antiquark going into a right-handed u antiquark and a W^+.

arrow or, alternatively, as a real number and a dial. The term shown in Figure 12.1b takes a right-handed d antiquark into a right-handed u antiquark and a W^+. General principles of quantum theory require that this coupling is proportional to V^*_{ud}. The star means that we take the arrow representing V_{ud} and reflect it about the horizontal axis. Said otherwise, the dial (phase) in V^*_{ud} is rotated by the same angle as the dial in V_{ud}, but in the opposite direction. The magnitudes of V_{ud} and V^*_{ud} (the sizes of the corresponding arrows) are equal. Thus, the two interactions in Figure 12.1 have the same intensity.

Charge conjugation changes the particles in Figure 12.1 into antiparticles, and vice versa. Parity transformation changes the spin (left into right, and vice versa). With a CP transformation, Figure 12.1a and b is interchanged. Thus, if CP were a good symmetry, then the couplings in the two figures should coincide and V_{ud} should coincide with V^*_{ud}. Because the arrows corresponding to V_{ud} and V^*_{ud} have opposite phases, these couplings can only coincide if their phases vanish. That is, if CP were conserved, then V_{ud} would be an ordinary real number. Couplings that are not real numbers are the only source of CP violation in the standard model.

However, many of the phases have no physical significance; for example, if we replace what we mean by a d quark by changing its phase, this will not matter. Only relative phases that do not change when you change the phase of a quark are significant; it then turns out that with only four quarks, there are no meaningful phases. With six quarks there are nine factors like V_{ud} forming a table with three rows and three columns known as the CKM matrix (as in Table 9.1). It turns out in general that this matrix can be determined by specifying three magnitudes and one phase.*

There are many ways to choose the four parameters that determine the CKM matrix. Once the hierarchy of magnitudes was determined by

---

* See Appendix 4 on unitary matrices.

TABLE 12.1   The Wolfenstein Parametrization of
the Cabibbo–Kobayashi–Maskawa (CKM) Matrix

V	d	s	b
u	$1-\lambda^2/2$	λ	$A\lambda^3 r$
c	$-\lambda$	$1-\lambda^2/2$	$A\lambda^2$
t	$A\lambda^3(1-r^*)$	$-A\lambda^2$	1

experiments (shown in Table 9.2), one of us (Lincoln Wolfenstein) proposed writing the matrix as shown in Table 12.1. This has become a standard way of writing the matrix. The crucial point is the use of the small
parameter λ (pronounced "lambda"). Experimentally, $\lambda = 0.22$, and terms
of order λ^4 (which is close to 0.0023) and smaller have been omitted. This
is sufficient for analyzing most experiments, although for some experiments the terms of λ^4 must be included. The bottom row, which gives the
coupling of the t quark, was determined by requirements on the internal
consistency of Table 12.1;* it shows that nearly all t quark decays go to B
mesons. The present fit to B decay rates gives A at about 0.8, and the magnitude of r is about 0.4. CP violation is included by giving r a phase γ (the
Greek letter gamma), while r^* has the same magnitude as r but the phase
$-\gamma$.† This unique phase is the only source of CP violation in the standard
model. Note that only two elements of the matrix have large phases: V_{ub}
with the phase γ and V_{td} with a phase denoted by β (pronounced "beta"
and, in fact, related to γ).

We now return to CP violation in the kaon system. The epsilon parameter, which can be blamed on CP violation in K0-K0bar mixing, is calculated
using the box diagram in Figure 11.2, where the quark q now includes the t
quark as well as the u and c quarks. The diagram with q set as the t quark
is proportional to $(V_{ts} V_{td}^*)^2$ and so includes the phase of $(V_{td}^*)^2$; this phase
is often referred to as 2β (pronounced "two times beta"). This diagram
with the t quark is greatly suppressed by λ^8 (approximately 0.0000055)
compared to the diagram with the c quark, which is proportional to λ^2
(approximately 0.05). However, the detailed calculation gives an enhance-
ment to the diagram with the t quark because of the large mass of the t
quark. While there are a number of uncertainties in the calculation, the
result is consistent with the experimental value $\varepsilon = 0.002$, provided the

---

* These requirements are known as unitarity constraints and are explained in Appendix 4.
† In mathematical terms, r is a complex number and r^* is the complex conjugate. r can be written as $\rho - i\eta$ and r^* as $\rho + i\eta$. The phase γ is given by $\tan\gamma = \eta/\rho$

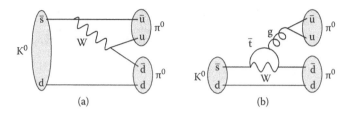

FIGURE 12.2   Diagrams leading to K0 decaying into π^0 π^0. (a) Dominant tree diagram; leads to $K_S \rightarrow \pi^0$ π^0 since it does not violate CP. (b) Penguin diagram needed in order to get K_L to decay. There is also a diagram where the gluon, g, is replaced by a photon.

FIGURE 12.3   How John Ellis tried to turn Figure 12.2b into a penguin.

phase 2β is fairly large. (The determination of β from other experiments is discussed in the next chapter.)

What about CP violation in the decay of kaons? The decay into $\pi^0\pi^0$ is mainly due to the direct transition where an s quark goes to a u quark, a u antiquark, and a d antiquark. This has no CP-violating phase and is shown in Figure 12.2a. For the K decays it is necessary to add to the dominant diagram a loop involving the t quark in order that the decay is sensitive to the phase in V_{td}, as shown in Figure 12.2b. For CP violation to occur in the decay, we need at least two diagrams with different phases, as happens in Figure 12.2. Note that the diagram in Figure 12.2b is proportional to λ^5, whereas the diagram in Figure 12.2a is proportional to λ, so that we expect the CP violation to be very small, as observed.

The loop diagram in Figure 12.2b is known as a penguin diagram because physicists have a keen sense of humor. The story has it that John Ellis (a theorist at CERN) and Melissa Franklin (now a Harvard experimentalist) were playing a game of darts at a pub and a bet was made. Having lost the game, Ellis was required to include the word *penguin* in his next scientific article. He realized that diagrams similar to the upper portion of Figure 12.2b can be deformed into Figure 12.3, which, if you have a really wild imagination, may seem to resemble a penguin. Thus, the term *penguin diagram* became part of the scientific jargon.

It turns out that the loop diagram in which the gluon is replaced by a photon is also important and tends to cancel the diagram with the gluon, so that the actual calculation of epsilon-prime is very difficult in the standard model. Within the great uncertainty of the calculation, the theoretical result is consistent with the experimental value. While it follows that CP violation in kaon decays can be explained by the standard model with three quark families, kaon physics cannot provide a quantitative test of the theory. This brings us to B mesons.

CHAPTER 13

CP Violation with B Mesons

13.1 CP VIOLATION AT THE B FACTORIES

There are two neutral B mesons. The B0 meson may be considered the combination of a d quark and a b antiquark. Applying the CP transformation we obtain a d antiquark and a b quark, which is known as the meson B0bar (pronounced "B-zero-bar"). As in the kaon system, because B0 and B0bar mix, they are not states with well-defined masses and lifetimes. The particles with well-defined masses are B_H (the heavier B meson) and B_L (the lightest B meson). We distinguish them by their mass and not by their lifetimes (as we did for the kaons) because their lifetime difference is very small. In the kaon system, K_L (L for long lived) lives much longer than K_S (S for short lived). By waiting long enough, a beam containing K_L and K_S will eventually contain almost only K_L; the K_S particles will have decayed away by then. This is not possible for the B mesons; fortunately, one can detect that B_H is heavier than B_L, although their mass difference is thirteen orders of magnitude smaller than their average mass. The reason is that the mass difference is comparable to the decay width, which in turn is related to the inverse of the lifetime.

In the standard model, the mixture is due to the so-called box diagram shown in Figure 13.1.[1] This diagram provides the main contribution to the mass difference between B_H and B_L. The fact that four weak vertices are required explains why the mass difference is so small (recall that each additional vertex implies a further power of the weak coupling constant

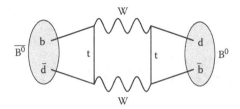

FIGURE 13.1   Dominant box diagram contribution to B-Bbar mixing.

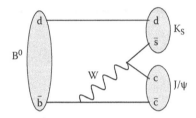

FIGURE 13.2   Tree-level diagram contributing to the decay B → J/ψ K_S. It is proportional to $V_{cb}^* V_{cs}$.

and a further suppression in perturbation theory). This contribution is proportional to $(V_{tb} V_{td}^*)^2$. In contrast to the kaon system, this diagram with a large phase from V_{td} dominates B0-B0bar mixing, and therefore, it predicts a large CP violation in mixing.

It turns out that the best way to detect this CP violation involves studying the time dependence of the transitions of B0 and B0bar to a final state f, which has a definite value (+1 or –1) of CP. Because of CP violation in the mixing, these transitions are not the same. The difference between them determines the phase of V_{td}, provided there is no CP violation in the decay process itself.

A particularly useful example is the final state J/Ψ K_S. The J/Ψ (pronounced "jay-psi") is the ccbar state with spin 1, already discussed in Chapter 6. K_S is the short-lived kaon, which is essentially the state K_+ (see Chapter 11). This state has CP = +1. The decay of B mesons into Ψ K_S is sometimes referred to as the gold-plated decay. This decay, when it occurs, is relatively easy to identify. The J/ψ is identified by its decay into a muon–antimuon pair, and the K_S is identified by its decay into two pions. These four charged particles are efficiently identified by current detectors.

In the standard model, the B → J/ψ K_S decay proceeds through the diagram in Figure 13.2. From the Cabibbo–Kobayashi–Maskawa matrix in Table 12.1, it is seen that there is no CP violation in the decay; that is, the

FIGURE 13.3   The full decay chain studied at the $\Upsilon(4S)$.

Cabibbo–Kobayashi–Maskawa matrix elements that appear in Figure 13.2 contain no phase. Therefore, the CP violation in the experiment described below is entirely due to the mixing given by Figure 13.1.

A clean way to produce neutral B mesons involves a particle known as the $\Upsilon(4S)$ (pronounced "upsilon-4-S"). This is a bound state of a b quark and a b antiquark that can be produced by smashing an electron with a positron at the relevant energy. Its mass is just large enough that it decays about half the time into a B0-B0bar pair and the other half of the time into a B+-B– pair. Figure 13.3 shows a typical experiment. Initially, the $\Upsilon(4S)$ decays into a B0-B0bar pair. As they evolve, the B0 can mix into a B0bar, and vice versa. At time t_1, the particle on the left-hand side has a decay identifying it as a B0 meson at that time (for example, due to the presence of a positron from its semileptonic decay). The beauty of quantum mechanics guarantees that the meson on the right-hand side must be a B0bar meson at that exact time.* Then, the B0bar on the right-hand side evolves in time, possibly remaining a B0bar, possibly mixing into a B0 meson. At time t_2, this meson decays into J/ψ K_S. By comparing this chain of events with its CP conjugate, one searches for CP violation.

There is one problem, however. If we sum all events (irrespective of the specific decay times), then we do not know whether t_1 preceded t_2 or events occurred the other way around. It turns out that the signal of interference CP violation drops out in such time-integrated analysis. These can only be used to look for direct CP violation. But there is no direct CP violation in the decays into J/ψ K_S. Thus, in order to detect CP violation in this decay, we must follow the time evolution. We must know the times.

---

* The $\Upsilon(4S)$ has spin equal to 1. Since the B mesons lack spin, conservation of angular momentum forces them to have a relative orbital angular momentum equal to 1. But this means that the corresponding wave function gets a minus sign under interchange of the two particles. If both particles were a B0 meson, then we would have two bosons in an antisymmetric wave function. This would be correct for fermions, but it is not possible for bosons. As a result, if at some instant we know that a B0 meson exists on one side, then at the same time, the other meson must be a B0bar. This uses as a tool the boson version of the experiment that Einstein, Podolsky, and Rosen found paradoxical for electrons. Paradoxes for the pioneers become technology in later years.

The time between one decay and the next is seen in the experiment as a distance between the positions of the two decays. Unfortunately, the B mesons arising from a $\Upsilon(4S)$ at rest only travel about 30 millionths of a meter in one lifetime. This distance is far too small to be measured. The solution was provided by Piermaria Oddone. He proposed that the colliding electron and positron beams should have different energies. The concept of asymmetric B factories had been born.

Two asymmetric B factories went into operation around the year 2000. One was at the Stanford Linear Accelerator Center (SLAC) in the United States, with a large detector named after a well-known character of children's books: the elephant Babar. The other was at the KEK laboratory in Japan with a detector called Belle.

At SLAC electrons were accelerated to an energy of 9 Gev and were collided with positrons having an energy of only 3 Gev. When the collision produces two B mesons, conservation of momentum requires that the B mesons are moving rapidly; in fact, their speed is approximately half the speed of light. On average, a B meson will move about a third of a millimeter before decaying. For the experiment comparing B0 and B0bar decays, it is necessary to measure the position of the point of decay accurately. In Babar, the decay points are located by the use of thin strips of silicon, which make it possible to locate the point to a few hundredths of a millimeter.

The decay that occurs first is identified as B0 or B0bar by a process known as tagging. For example, if a positive electron or positive muon is identified, this means that the original decay came from a b antiquark (such as the decay where the b antiquark goes into a c antiquark, a positron, and a neutrino) and so from a B0. Then, for the second decay, one looks for the CP-even state, like $J/\Psi\ K_S$ identified by the two muons from the J/Ψ decay and the two pions from the K_S decay. This is the sequence illustrated in Figure 13.3. What one measures is a difference between the rate when the tagged particle is a B0 and the rate when the tagged particle is a B0bar, as illustrated in Figure 13.4. What is plotted is the difference between the two rates divided by the sum as a function of the time difference. This is called the asymmetry. The time dependence results from the time required for B0bar to become a B0, or vice versa. This depends on the magnitude of the mixing, as given by Figure 13.1. So, the asymmetry depends on the CP-violating phase of Figure 13.1, which is the phase of V_{td} squared, denoted as 2β. (Remember that β is the phase of V_{td} in Table 12.1.)

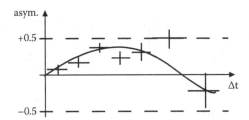

FIGURE 13.4 Sketch of the CP-violating decay asymmetry in the decay of B mesons into J/ψ K_S. The crosses represent experimental points with their respective errors; the solid curve represents the fit to the theoretical prediction. Here, Δt is the time difference $t_2 - t_1$ as seen in Figure 13.3. (Sketch based on B. Aubert et. al., Babar Collaboration, Physical Review Letters 94, 161803 (2005), fig 2b), pg 161803–6)

We have noted in the case of the kaons that there can be a large uncertainty in the theoretical calculation using the standard model. The reason is primarily that the calculations are done for quarks, as shown in the various diagrams, and there may be large theoretical problems in going from quarks to mesons. This is not a problem in determining 2β. The reason is that 2β is determined from the ratio of the difference in rates to the sum of the rates, so that the uncertain factor cancels out. Alternatively, we can say that the asymmetry for the mesons is the same as for the quarks.

Although the accelerator and detector in Japan have many technical differences, the basic idea is the same. Much confidence has been gained by comparing the results of the two experiments and combining them. Thus, the sine of 2β has been determined to equal 0.68 with an experimental uncertainty of only 0.04. This sets 2β to be 43°, with an uncertainty of 2°. The importance of this result will be seen in Chapter 14.

This CP violation is analogous to epsilon in the K0 system in that it can be attributed to CP violation in the B0-B0bar mixing. What about CP violation that is clearly in the decay, analogous to epsilon prime in the K0 system? This was first discovered in the observation of the difference between the decays B0 → K^+ π^- and B0bar → K^- π^+, which was found to be a 10% effect. Recall that two diagrams with different phases are required for CP violation in a decay. The two contributions to this decay are shown in Figure 13.5. The direct diagram in Figure 13.5a, or tree diagram, is proportional to $V_{ub}^* V_{us}$ and has a phase known as –γ (pronounced "minus gamma"), while the penguin diagram in Figure 13.5b is proportional to $V_{tb}^* V_{ts}$ and so has no phase. It is the interference between these two terms that is responsible for the direct CP violation. Although it is generally

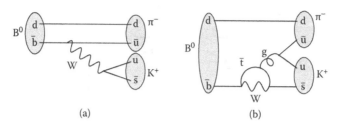

FIGURE 13.5 Two contributions to the decay B0 → K⁻ π⁺. The first diagram is proportional to $V_{ub}^* V_{us}$, and the second is proportional to $V_{tb}^* V_{ts}$. Notice the presence of the gluon (g) in the second diagram.

expected that the tree diagram should be very dominant, in this case the tree diagram is suppressed by a factor λ^4 (approximately 0.002), while the penguin diagram has a factor λ^2 (approximately 0.05). As a result, the two contributions are comparable, allowing for a sizable direct CP violation. This is in contrast to the K0 system where the penguin diagram is highly suppressed. As for the case of the K0 system, it is not possible to do a quantitative theoretical calculation of this asymmetry in the standard model, and so, while the result is consistent with the standard model qualitatively, it cannot be used to determine gamma.

13.2 CP VIOLATION IN THE BS SYSTEM

In addition to the B0, which consists of a b antiquark combined with a d quark, there is a Bs meson, which consists of a b antiquark combined with an s quark. Similarly, there is a Bsbar made of a b quark combined with an s antiquark. As in the case of the B mesons, Bs and Bsbar mix so that the particles with defined masses Bs_H (the heavier) and Bs_L (the lighter) are mixtures of Bs and Bsbar. Unlike the case of the B mesons, there is a measurable lifetime difference between Bs_H and Bs_L, with Bs_H having the longer lifetime.

In the standard model, the mass difference and mixing are due to the box diagram of Figure 13.6.[2] This contribution is proportional to $(V_{tb} V_{ts}^*)^2$. The form of the Cabibbo–Kobayashi–Maskawa matrix in Table 12.1 tells us that there are two important differences compared to the B0 system. First, the mass difference is much larger by a factor of order $1/\lambda^2$ (approximately 21). Second, there is no CP-violating phase. If one goes beyond the approximation in Table 12.1, there is actually a small phase proportional to λ^2; it is approximately equal to 0.04.[3]

A major goal for the future is to test the standard model of CP violation and to search for possible new sources of CP violation. It is thought that the mixing, which is a second-order effect in the standard model, might be particularly sensitive to new physics. A possibility is to try to carry out an experiment analogous to that which measured the sine of 2β replacing the B0 and B0bar by Bs and Bsbar. The idea is to check whether the CP-violating effect is very small, as predicted by the standard model.

In the case of B, the gold-plated decay was to J/Ψ Ks, a CP state with well-defined CP properties. In the case of Bs one can consider the decay to the state J/ψ Φ (pronounced "jay-psi-phi"). As seen before, J/ψ is made up of a c quark and a c antiquark. Φ is essentially made up of an s quark and an s antiquark. The corresponding decay diagram is shown in Figure 13.7. This is not actually a state with definite CP properties, but theory and experiment suggest it is primarily a CP = +1 state.* The CP-violating asymmetry in this decay provides a clean measurement of the phase between the V_{cb}^* V_{cs} factor from the decay in Figure 13.7 and the V_{tb}^* V_{ts} factor from the mixing in Figure 13.6. As in the case of B → J/Ψ Ks, the decay is proportional to V_{cb}^* V_{cs} and so has no CP-violating phase. One can then measure the decay of Bs → J/Ψ Φ as a function of time and compare its CP conjugate, Bsbar → J/Ψ Φ. The difference measures the analog of the sine of 2β

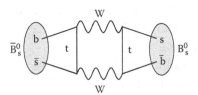

FIGURE 13.6   Dominant box diagram contribution to B_s-$B_s$bar mixing.

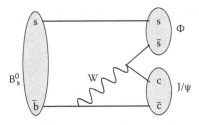

FIGURE 13.7   Tree-level diagram contributing to the decay B_s → J/ψ Φ. It is proportional to V_{cb}^* V_{cs}.

---

* This means that it also has a very small component with CP = −1.

for the B system. In this case we expect a very small asymmetry between the Bs and Bsbar decays since there is no CP violation in the decay and only the very small CP violation in the mixing of order 0.04.

In the case of the B mesons it was possible to produce an abundance of B0-B0bar pairs at the asymmetric B factory colliding an electron and a positron at a particular energy. There is no such wonderful source for Bs-Bsbar pairs. Thus, the experiment on Bs mesons must be carried out at high-energy proton-proton colliders, which produce a multitude of particles, including Bs and Bsbar mesons. The first experiments were carried out at Fermilab. In the production, one always produces a b quark together with a b antiquark. Thus, it is determined whether the original meson is Bs or Bsbar by "tagging," that is, detecting whether the particle produced with it contains a b quark or a b antiquark. For example, the particle produced with the neutral meson of interest might be a B⁺ (decaying to J/Ψ K⁺). But B⁺ involves a b antiquark, so that one must have had a Bsbar meson as a partner to start with. (Remember that the Bsbar meson is the one with a b quark, while the one with a b antiquark quark is named Bs.) The first results from Fermilab have shown an asymmetry significantly larger than expected. However, given the experimental errors, it is too soon to claim a clear signal of new physics.

There are a number of other decays that might be used. An interesting example is the decay Bs to $D_s^+ D_s^-$, based on the diagram in Figure 13.8. This is a final state with CP = +1.

The Large Hadron Collider (LHC) is the newest proton-proton collider, beginning operation at CERN in Geneva. While a major goal is the production of new particles like the Higgs boson (see Part D of the book), there is a special detector called LHCb designed to study B and Bs mesons with greater precision than has been possible so far. In particular, there is

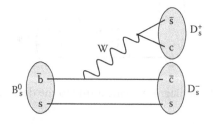

FIGURE 13.8   Tree-level diagram contributing to the decay $B_s \rightarrow D_s^+ D_s^-$. It is also proportional to $V_{cb}^* V_{cs}$.

the hope to measure the CP-violating asymmetry in Bs, Bsbar decays to an accuracy of 0.02, and thus determine whether it is significantly larger than the small value of 0.04 expected in the standard model.

NOTES

1. There are in fact nine box diagrams, where the two charge +2/3 internal quarks appear in the combinations uu, uc, ut, cu, cc, ct, tu, tc, and tt. However, explicit calculations show that by far the largest contribution comes from the tt diagram shown in Figure 13.1 because of the large top quark mass. Note that, unlike the kaon case, the tt diagram is not suppressed by the Cabibbo–Kobayashi–Maskawa matrix elements; in fact, all nine diagrams have Cabibbo–Kobayashi–Maskawa coefficients proportional to lambda to the sixth power.
2. There are in fact nine box diagrams, where the two charge +2/3 internal quarks appear in the combinations uu, uc, ut, cu, cc, ct, tu, tc, and tt. But, as in the case of B0, the tt diagram shown in Figure 17.1 dominates.
3. From the approximate form of the Cabibbo–Kobayashi–Maskawa matrix in Table 12.1 you find no CP violation. However, there are corrections of order $2\,\eta\,\lambda^2 = 0.04$ that lead to CP violation.

CHAPTER 14

Checking the Standard Model

The Rho-Eta Plane

CP violation enters the standard model with the addition of the third family of quarks (b and t). This was noticed by Makoto Kobayashi and Toshihide Maskawa when only two quark families were known. The phase that determines CP violation is found in the elements of the Cabibbo–Kobayashi–Maskawa matrix V_{ub} and V_{td}. In the parametrization of the matrix given in Table 12.1, two of the parameters are known with good precision, and we can fix them as $\lambda = 0.226$ and $A = 0.82$. The remaining two parameters enter only in V_{ub} and V_{td} and are given by the magnitude and the phase of V_{ub} or the parameter r in Table 12.1.

We recall that r corresponds to an arrow on the (so-called complex) plane. We can view it as a number (magnitude) and a dial (phase). The arrow r* has the same magnitude as r, but the opposite phase. Alternatively, we may consider the horizontal and vertical components of the arrow r, as illustrated in Figure 14.1. These are designated by the Greek letters ρ (pronounced "rho") and η (pronounced "eta"). The goal of experiments is to determine the values of ρ and η in different ways to test whether the measurements are consistent or whether some physics beyond the standard model is needed.

In Figure 14.2, we show the Cabibbo–Kobayashi–Maskawa matrix elements $(V_{ub}/A\lambda^3)$ and $(V_{td}/A\lambda^3)$ in the ρ-η plane. Of course, $V_{ub}/A\lambda^3$ is what we defined to be r. In the figure, we also indicate the phases γ and β, which determine the way in which V_{ub} and V_{td} violate the CP invariance.

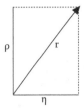

FIGURE 14.1 The arrow r and its horizontal and vertical components (ρ and η, respectively).

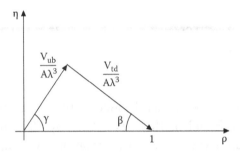

FIGURE 14.2 The unitarity triangle.

As discussed in Appendix 4, Figure 14.2 is a version of what is called the unitarity triangle.

One way to determine ρ and η is to measure the sides of the triangle in Figure 14.2, that is, to determine the magnitudes of V_{ub} and V_{td}. The magnitude of V_{ub} determines the rate of decays arising from b → u transitions such as B meson decaying into a pion, an electron, and a neutrino. There is considerable uncertainty in going from the theoretical b → u rate to the experimental B to pion rate, so that this measure only determines a range of values for the magnitude of r, given by the square root of $\rho^2 + \eta^2$. For example, if the experiments determine r to lie between 0.3 and 0.5, then the constraint in the ρ-η plane is illustrated by the circular region in Figure 14.3.

The magnitude of V_{td} can be determined in principle from Δm_d, the mass difference between the two neutral B mesons, B_H and B_L, as determined from the box diagram of Figure 13.1. There is a large uncertainty from hadronization, that is, from going from the four quarks sticking out of the box diagram into the observed B mesons, as shown in Figure 13.1. It turns out that the study of the corresponding mass difference in the Bs system proved to be a great help.

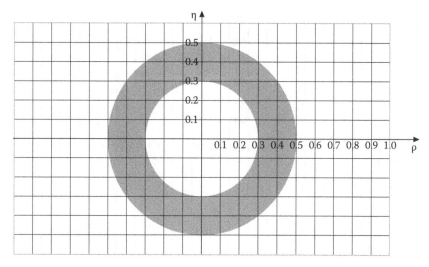

FIGURE 14.3   Constraint in the rho-eta plane from a measurement of the magnitude of V_{ub}.

The only difference between the Bs-Bsbar mixing of Figure 14.1 and the B0-B0bar mixing of Figure 13.1 is the exchange of s by d. Each hadronization mechanism is largely unknown. We mentioned that B0-B0bar mixing, involving b and d quarks, allowed a rather imprecise probe of V_{td}. Similarly, the prediction for Bs-Bsbar mixing is imprecise. Fortunately, the comparison between the two is much more precise because only the difference between s and d quarks comes into play. As a result, comparing the mass differences in the two systems allows a better determination of V_{td} than that obtained from B0-B0bar mixing alone. It is curious that a measurement of Bs-Bsbar mixing, which is not directly related to V_{td}, improves the precision of the determination of V_{td} by a factor of 3 just because it provided a better handle on the hadronic uncertainties. As in real life, help may come from unexpected places.

The experimental measurement of the magnitude of V_{td} determines a circle centered at value $\rho = 1$, $\eta = 0$. For example, the measurements may determine that the magnitude of $V_{td}/A\lambda^3$ lies between 0.8 and 1.0. Combined with V_{ub}, this would lead to Figure 14.4. The intersection of the two curves provides important constraints on ρ and η, assuming that the standard model holds. Although these experiments do not involve any CP-violating observables, one sees from Figure 14.4 that the phases of V_{ub} and V_{td} are distinctively different from zero. As shown in Figure 14.5,

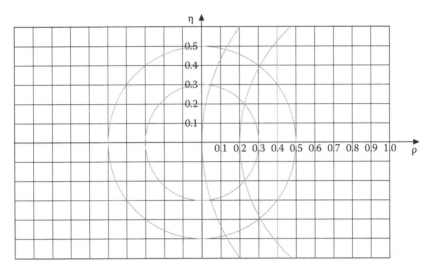

FIGURE 14.4   Combined constraints in the rho-eta plane from the magnitudes of V_{ub} and V_{td}.

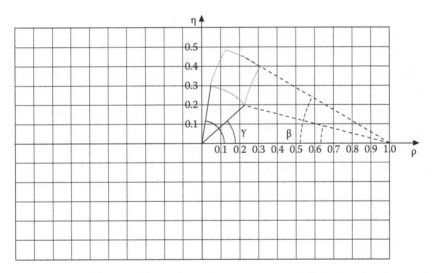

FIGURE 14.5   The curved lines show the extreme possibilities for the phases of V_{ub} (solid line, γ) and V_{td} (dashed line, β), consistent with Figure 14.4.

given the constraints from Figure 14.4, there is a limited range for the phases γ and β.

As discussed in Chapter 13, the observation of CP violation in time-dependent decays of B0 and B0bar to J/Ψ K_S provided a determination of the sine of twice the angle β, with practically no theoretical uncertainty.

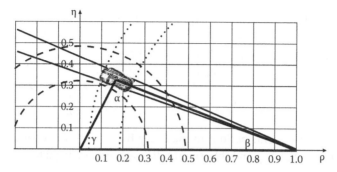

FIGURE 14.6   Sketch of the combination of the most important constraints on the rho-eta plane available by the end of 2007. The constraints come from V_{ub} (dashed line), V_{td} (dotted line) obtained from mixing in the B and Bs systems, and β from CP violation in B → J/ψ K_S. Once the statistics of these and other experiments are taken into account, one obtains the dark peanut-shaped allowed region. Also shown are the angles α, β, and γ of the unitarity triangle.

The constraints on β from the magnitude of V_{ub} and V_{td} are completely consistent with that determination. This can be considered a success of the standard model. The results available by 2007 on the major constraints on the ρ-η plane are shown in Figure 14.6, inspired by the CKMFitter group.[1] Notice that all results are consistent if ρ and η lie in the small dark peanut-shaped region. This means that the standard model survived a large number of tests with flying colors. The combined results give us a very small permitted region in the rho-eta plane, around ρ = 0.14 and η = 0.34. As a result, we now know all elements of the Cabibbo–Kobayashi–Maskawa matrix with fairly good precision.

It is hoped that, with future experiments and theoretical calculations, the results shown in Figure 14.6 can be more precise so that small deviations from the standard model might be detected. Also, it is expected that additional experimental quantities will be measured with sufficient precision to further constrain the values of ρ and η.

An important example would be a direct determination of the phase γ of V_{ub}. In principle, this should be possible by finding CP violation in a decay involving a b quark going into a u quark. An example could be B → π⁺ π⁻, which one expects to be due to Figure 14.7a. If we then study the difference between B0 → π⁺ π⁻ and B0bar → π⁺ π⁻ as a function of time, the result will involve CP violation in the mixing (as in the case of the decays into J/Ψ K_S discussed in Chapter 13) plus the violation of CP in the decay, due to the phase γ. Thus, the difference in time-dependent decay rates

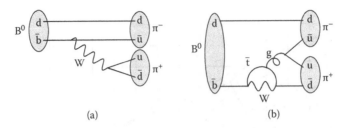

FIGURE 14.7 Two contributions to the decay B → π^- π^+: (a) tree-level diagram, proportional to $V_{ub}^* V_{ud}$, and (b) penguin diagram, proportional to $V_{tb}^* V_{td}$.

would measure the sine of twice (β + γ). Given β, this would be a method to determine γ.

However, there is a problem. A significant contribution to the decay can come from the penguin diagram of Figure 14.7b. This was originally referred to as penguin pollution because it meant that the phase of the decay amplitude was not γ, so that this method to determine γ was not reliable. A variety of experiments on the decays to two pions and to two rho-mesons (heavier versions of the pion) have put limits on the amount of penguin pollution. The conclusion is that (β + γ) can be determined from these CP violation measurements with an uncertainty around ±20°, and that γ is around 60°. Again, this is in agreement with the results shown in Figures 14.5 and 14.6.

The direct CP violation in decay B → K π, due to the diagrams in Figure 13.5, also depends on γ. However, the theoretical uncertainties are too great to allow for a determination of γ, but the large value of γ from Figure 14.6 is consistent with that observed in these decays. Similarly, the value of the parameter ε measuring CP violation in K0-K0bar mixing depends on β and the magnitude of V_{td}. Within the theoretical uncertainties, the experimental value is consistent with the values in Figure 14.6.

NOTE

1. http://ckmfitter.in2p3.fr/.

CHAPTER 15

CP Violation

Where Do We Go from Here?

T HE PRESENT RESULTS SUMMARIZED in Figure 14.6 provide strong evidence for the CP-violating phase required by the Cabibbo–Kobayashi–Maskawa (CKM) model. The question that remains is whether there are other sources of CP violation not yet discovered. The CP violation from the CKM matrix is not sufficient to explain how the matter-antimatter asymmetry of the universe was created.

There are two general ways to attack this question: (1) to gather more precise constraints on the rho-eta plane, or (2) to look for CP-violating effects in cases where the standard model predicts very small or no effects.

15.1 FURTHER CONSTRAINTS ON THE RHO-ETA PLANE

An example of a very interesting decay that has not yet been detected is the decay of a K_L into a neutral pion, a neutrino, and an antineutrino. Since this final state is CP even, the decay from K$_-$ is CP violating.* The decay is calculated from the diagrams in Figure 15.1. The decay amplitude is directly proportional to the CP-violating part of $V_{td} V_{ts}$,* which is given by $A^2 \lambda^5 \eta$. Unlike other decays, in this case, the rate can be calculated quite accurately in terms of the value of η. As the value of η is determined from B decays (as in Figure 14.6), this could provide a good test of the standard model or a wonderful way to detect new physics. Given our present knowledge of η, it is predicted that only 3 out of 100 billion K_L decays will go to this final state, so that the experiment will be very difficult. On the other hand, this small rate means that a new interaction that violates CP

---

* The small K_+ mixed into K_L is not significant.

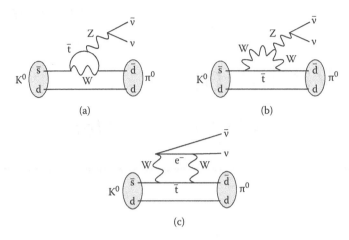

FIGURE 15.1 Diagrams leading to K_L going to π^0, a neutrino, and an anti-neutrino. Note that the diagrams involve the weak interaction twice, and each involves $V_{td} \times V_{ts}$.

and allows s to go directly to d, a neutrino, and an antineutrino could be almost a million times weaker than the standard model weak interaction and still have a significant effect on the rate.

Experiments to measure this decay were proposed at both Fermilab and Brookhaven in the United States, but neither received funding. An experiment is now being planned at the new accelerator JPARC in Japan. A high-intensity proton beam hitting a target will produce a billion K_L mesons every 3 seconds. These will enter a vacuum chamber surrounded by photon detectors, and the products of the K_L decay in flight will be detected. The signal for the decay would be two photons from the decay of the neutral pion and nothing else. The "nothing else" would be the two neutrinos. The ultimate goal is to detect one thousand of these decays in a year. The greatest problem is to avoid a fake signal; even the rare decay into $\pi^0 \pi^0$ (which was the subject of the experiment that measured epsilon-prime) might give a signal if two of the four photons are not detected. A number of other rare kaon decays will be studied as well.

Most of the results on B decays come from the electron-positron colliders at the SLAC laboratory in the United States and KEK in Japan (known as B factories). Although the laboratory in the United States is closing down, it is planned to increase intensity and obtain more data at the B factory in Japan. The hope is that additional experiments combined with new theoretical calculations can decrease the uncertainty of the results

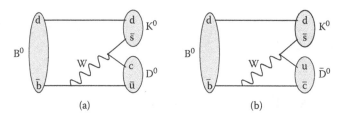

FIGURE 15.2 Simple diagrams driving the decay: (a) B0 → D0 K0 and (b) B0 → D0bar K0.

summarized in Figure 14.6. A major goal remains to directly determine the phase γ of V_{ub}.

A promising method to determine γ is derived from the original proposal of Michael Gronau, Daniel Wyler, and David London. It uses the decay of B mesons to D mesons plus K mesons. The basic idea is illustrated by the diagrams in Figure 15.2, describing B0 going to D0 (D0bar) plus K0. The diagram in Figure 15.2a has the phase γ and goes to D0 K0, while the diagram in Figure 15.2b has no CP-violating phase and goes to D0bar K0. By detecting the characteristic decays of D0 and D0bar, the rates of each of these decays can be measured. For example, a D0 decays into $K^- \pi^+$, while a D0bar decays into a $K^+ + \pi^-$. Then, it is also possible to look at the decays into a state of definite CP properties, like $K_S \pi^0$. This means that the final state contains an equal mixture of D0 and D0bar, and therefore that the amplitude for this decay is obtained by adding the amplitudes of Figure 15.2a and b. The result will depend on the relative phase of the two diagrams, which is gamma. Thus, the measurement of the rate going to $K_S \pi^0$ gives information on the phase gamma, even though no CP violation has been observed. It is also possible to repeat this for the B0bar, which would involve CP violation and also help to determine gamma. There are many variations of this idea. It is hoped that gamma will be determined from B decays into DK with an uncertainty of only 2°. Complementary results will be obtained at the Large Hadron Collider (LHC) with the LHCb detector.

15.2 QUANTITIES THAT ARE SMALL IN THE STANDARD MODEL

An interesting way to search for new physics is to measure observables that are very small in the standard model. There is the possibility that the characteristics of the standard model that made these quantities small do not hold for the new physics. Thus, even though the new physics interactions

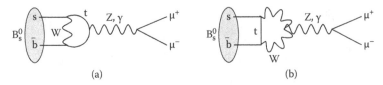

FIGURE 15.3 Standard model diagrams responsible for the decay B → μ⁺ μ⁻.

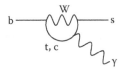

FIGURE 15.4 Standard model diagram responsible for the decay b → s γ.

may be generally smaller than the standard model, they may show up in these observables. An example discussed in Chapter 14 is the asymmetry in Bs-Bsbar mixing analogous to β (beta) in the B0-B0bar system.

Also of interest at LHCb are rare decays of the Bs, such as the decay into a muon-antimuon pair, which has never been observed. In the standard model, this occurs via the loop diagrams shown in Figure 15.3. This gives a branching ratio of about 3×10^{-9}. Because it arises from complicated loop diagrams, there is the possibility of significant contributions from new heavy particles, if they exist.

Also of interest are decays in which a b quark decays into an s quark and a photon. This decay rate has been measured with an accuracy of about 10%, and the rate agrees with the theory from the diagrams like those of Figure 15.4. In the standard model, there is negligible CP violation; that is, the rate for the b quark decaying into an s quark and a photon is the same as the rate for a b antiquark to decay into an s antiquark and a photon. The observation of a significant CP violation would be evidence of a CP-violating amplitude interfering with that of the standard model.

The D0 meson has a c quark and a u antiquark, while D0bar has a c antiquark and a u quark. Recently, a small D0-D0bar mixing has been observed, analogous to the mixing of K0-K0bar or B0-B0bar. This is presumed to be due to Figure 15.5 with intermediate s or d quarks. Very little CP violation in the mixing is expected, because this would require intermediate b quarks and so would be proportional to $(V_{cb} V_{ub})^2$, which is proportional to λ^{10}, that is, of order 10^{-7}. This should be compared to the case of s and d quarks, which is proportional to λ^2, that is, to about 4×10^{-2}.

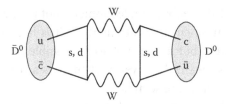

FIGURE 15.5   Important standard model contribution to D-Dbar mixing.

Thus, the observation of CP violation at the level greater than 1% would be a signal of new physics. This again is a case where a small new physics contribution can be detected if it violates CP invariance.

There is a totally different method to explore possible new physics that violates CP invariance: it involves looking for violations of time reversal invariance T. The point is that any theory we consider has CPT invariance, so that if it violates CP it will also violate T. However, it is very difficult to directly test T invariance with the types of experiments we have discussed so far. You cannot really reverse the process of weak decays, and you cannot reverse a high-energy collision. So we turn to experiments in low-energy nuclear and atomic physics.

The key is to search for electric dipole moments of the electron or neutron. These particles have spin, as discussed in Chapter 3, and like any spinning charge they have a magnetic moment.* This means they interact with an external magnetic field so that changing the direction of the spin from parallel to the field to antiparallel changes their energy. If their spin is fixed to be perpendicular to the field, the spin will rotate around the field, a process also known as precession.

However, T invariance says that there is no such effect in an electric field. The point is that reversing time reverses the direction of the spin and also reverses the direction of the magnetic field. (Remember that magnetic fields are due to electric currents.) But the electric field due to stationary charges does not change under time reversal. A nonzero value of the electric moment would thus be a signal of T invariance violation, and arise from a CP-violating interaction.

In the standard model a neutron electric dipole moment can arise only from very complicated diagrams. The result is hard to calculate, but it is approximately 10^{-32} in appropriate units (e-cm). This is 18 orders of magnitude smaller than the magnetic moment. A number of experiments have

---

* Although the neutron has no charge, it is made up of u and d quarks that do have charges.

tried to measure this electric dipole moment, but the best that can be said today is that the moment is less than 10^{-26}. It is hoped that new experiments can do better, but it does not seem possible to reach the standard model value. This means, however, that there is a real possibility to discover new physics in future searches.

The two experiments that provide the present limit used neutrons from nuclear reactors, one in Grenoble, France, and one in St. Petersburg (formerly Leningrad), Russia. The neutrons are slowed down to a very low speed by sending them through suitable materials. By the use of magnetic material, the neutron spin is polarized; that is, all the neutrons have their spin pointing in one direction. The neutrons are then trapped in a large cylindrical vessel for a short time; it is possible to trap the neutrons because they are reflected from the walls of the vessel. A magnetic field is turned on in a direction perpendicular to the neutron spin. As a result of the magnetic moment, the spin precesses. By measuring the direction of the spin when the neutrons leave the vessel, the rate at which the neutron spin was rotating can be measured. Then a strong electric field is added, and it is seen whether the rate of precession is changed, as it would be by an electric dipole moment. The experiment must be done very carefully to avoid small uncertainties in the magnetic field; up to now no effect has been seen.

A new experiment is planned with the goal of improving the sensitivity by a factor of 100. The plan has been developed over a period of more than 10 years, and it is hoped that the experiment can start in the near future. The neutrons will come from the bombardment of a target with protons at the Neutron Spallation Source at Oak Ridge Laboratory in the United States. The major new idea is to fill the vessel with liquid helium; helium becomes a liquid only at 4° above absolute zero. It is expected that many more neutrons can be stored and that they will stay longer. There are other measures involved to control the magnetic field.

In the standard model, the electric dipole moment of the electron is much smaller than that of the neutron and much too small to ever measure. Again, this is a great way to look for new physics. One cannot measure the moment for a free electron because interaction with its electric charge would ruin any experiment. The key is to measure the electric dipole moment of an atom or a molecule, and then use calculations from atomic physics to relate this to the electric dipole moment of an electron inside. It turns out that, in some cases, the theory says that the electric dipole moment of the atom or molecule is many times larger than that of the electron, which greatly enhances the sensitivity of the experiment. The present limit on this moment is more than ten times better than that on the neutron.

PART C

The Amazing Story of the Neutrino

It took 25 years from Pauli's theoretical suggestion until the first detection of the neutrino in 1955. The second neutrino, the muon neutrino, was detected in 1962. By 1970, 40 years after Pauli, that was all: two detections of this elusive particle. However, in the following 30 years experiments involving neutrinos became a major part of elementary particle research and of importance for astrophysics. Once the excitement of neutrino physics was realized, the original detections were awarded Nobel Prizes: Schwartz, Lederman, and Steinberger in 1988 for the muon neutrino (ν_μ) and Fred Reines in 1995 for the electron neutrino (ν_e).

As discussed in Chapter 8, an experiment at the CERN accelerator in 1973 showing the elastic scattering of muon neutrinos from neutrons and electrons provided the first evidence for neutral current interactions, an essential feature of the standard model of weak interactions. Neutrino physics advanced slowly because neutrinos interacted weakly, and so were hard to detect. On the other hand, because they had no other interactions, neutrinos at CERN provided a wonderful way to learn about weak interactions.

This early history of neutrinos depended on man-made sources, nuclear reactors, or particle accelerators. The next big advance had to do with neutrinos provided by nature, atmospheric neutrinos and solar neutrinos. The earth is bombarded from outer space with high-energy particles, mainly protons, called cosmic rays. When these protons hit the nuclei of oxygen or nitrogen they produce pions, and the pions decay into muons and muon neutrinos. Each muon decay yields one muon neutrino and one electron neutrino. Overall, the complete pion decay chain produces one electron neutrino and two muon neutrinos. These are the atmospheric neutrinos. Stars are powered by nuclear energy; as a result of these nuclear processes, some of the energy is emitted as neutrinos. Whenever you see the light from a star, there is likely to be a flux of neutrinos coming with it. The neutrinos from the sun are solar neutrinos.

It was from studying these neutrinos that we first had evidence that neutrinos had a mass. In the standard model neutrinos were massless, and so this provides the first evidence for physics beyond the standard model. We do not actually know the masses, only that there are three different masses and that all are very small, more than a million times smaller than the mass of the electron.

CHAPTER 16

The Mystery of the Missing Neutrinos

Neutrino Oscillations

THE SYSTEMATIC STUDY OF atmospheric neutrinos became possible with the development of large-water Cerenkov detectors. When a charged particle moves with high energy through water, it produces along its path a cascade of photons, an effect originally discovered by Pavlov Cerenkov in 1934.[1] The detectors consist of a large tank of water surrounded by photomultipliers that detect the photons, making it possible to trace the path of the particle (Figure 16.1). The goal is to see electrons or muons whose path begins in the middle of the detector, indicating that they were produced as a result of collisions of incident neutrinos with nuclei of hydrogen or oxygen. Of great importance is the possibility of distinguishing muons from electrons. This is possible because the electrons, being much lighter, scatter in the water and so provide a quite different signal in the photon detectors.

In the 1980s two such detectors were brought into operation: Kamiokande in the Kamioka mine in Japan, and IMB in a salt mine in the United States. The original motivation for these detectors was a search for the decay of the proton. While they never saw proton decay, they began a very exciting time for neutrino physics. In observing the atmospheric neutrinos, they found that the ratio of muon neutrino to electron neutrino appeared to be less than 2 to 1. Recall that the ratio 2 to 1 was expected since the pion decay and the muon decay each pro-

FIGURE 16.1 Cerenkov radiation from an electron moving through water. The dashed line represents the electron path. The wiggly lines are emitted photons.

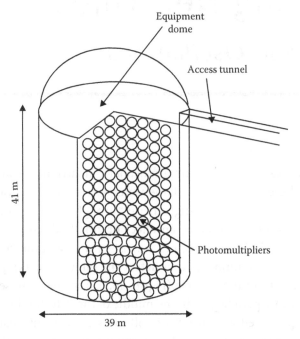

FIGURE 16.2 Scheme of the SuperKamiokande experiment. The large cylinder is filled with water and surrounded by photomultipliers.

duce one muon neutrino, while there is only one electron neutrino from muon decay.

The definitive evidence came when a much larger detector, SuperKamiokande, came into operation containing 50,000 tons of water viewed by 11,000 20-in. photomultipliers, as shown in Figure 16.2. The atmospheric neutrinos bombard all parts of the earth. The neutrino detector will detect neutrinos coming down from above, but also neutrinos from the other side of the earth coming up. Knowing how weakly neutrinos interact, we expect less than one in a thousand will be stopped on their way through the earth. Just as many neutrinos should be seen coming up as coming down. The results are shown in Figure 16.3. For the muon neutrinos only half as many are seen coming up!

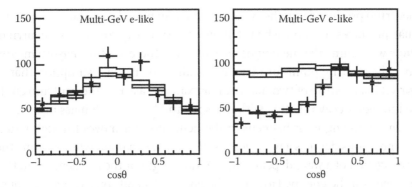

FIGURE 16.3 Angular distribution of the atmospheric neutrinos detected by SuperKamiokande, identified as electron neutrinos or muon neutrinos. Cosθ = +1 is downward and cosθ = −1 is upward. Black dots with error bars show the data in different angular distribution bins. The boxes show the results that would be expected if there were no oscillations. Solid lines show the fit assuming oscillations. (From A Measurement of Atmospheric Neutrino Oscillation Parameters by Super-Kamiokande 1, The Super Kamiokande Collaboration, Phys. Rev. D71, 112005–112023, Fig. 31 [first two figures on the third line of figures], [hep-ex/0501464]. With permission.)

What could be the explanation? There was no possible neutrino destroyer hidden in the center of the earth. It was known from observations that the cosmic ray intensity was the same on the other side of the earth. The conclusion was that half of the muon neutrinos had transformed to another type of neutrino as they traveled the 8,000 miles through the earth. Since there was no excess of upward-going electron neutrinos, it was concluded that they had transformed into the third type, the tau neutrino. Unfortunately, the tau neutrino is very difficult to detect, and so only very limited evidence could be produced for upward-going tau neutrinos. Nevertheless, that was the conclusion reached by a process of elimination. The possibility that neutrinos could be transformed, a process called neutrino oscillation, was first proposed by Bruno Pontecorvo, and detailed in 1962 by Maki, Nakagawa, and Sakata. It meant that neutrinos had a mass! For many years the idea of neutrino oscillations seemed to be just a random theoretical speculation; then gradually it became clear that this was the solution to the mystery of the disappearing neutrinos.

Let us for simplicity consider just two types of neutrinos, say, muon neutrino (v_μ) and tau neutrino (v_τ). When we say muon neutrino we are defining the neutrino by the way it interacts. Thus, muon neutrino is the neutrino produced together with the muon in pion decay and the neutrino that produces a muon when it interacts. Similarly for the tau neutrino. Now we assume that neutrinos have mass, so there are two masses: m_1 and m_2. The starting point for neutrino oscillations is the assumption that v_μ (or v_τ) is a mixture of two masses. For example, assume the neutrino emitted in pion decay is half v_1 with mass m_1 and half v_2 with mass m_2.

In discussing what happens to the neutrino as it moves through space, we must consider it as a wave, just as in the case of a photon or electron. The propagation of a wave depends on the wavelength or the frequency, and this depends on the energy. However, the energy depends on the mass because mass contributes the term mc^2 to the energy. Thus, as the neutrino moves, the v_1 and v_2 components get out of phase, as shown in Figure 16.4. What we observe as v_μ is the mixture of v_1 and v_2 in phase; when v_1 and v_2 have opposite phases this is no longer v_μ. In the simple case of only two neutrinos this is v_τ. The distance the neutrino has to travel until the v_μ transforms (or oscillates) into v_τ and back again into v_μ is called the oscillation length.

It is clear that the smaller the mass difference, the larger the oscillation length and the longer the distance the neutrino has to travel for the oscillation to be detected. Thus, the atmospheric neutrinos coming down have traveled on average 15 km and have not oscillated significantly, but those coming upward have traveled 8,000 km. Original attempts to see these oscillations with v_μ from accelerators failed because the detectors were much too near. The oscillation length also increases with increased energy because it depends on the ratio of the mass difference to the energy. As an example, if $m_2 = 0.05$ eV (10 million times smaller than the mass of the electron) and $m_1 = 0$, then the oscillation length is about 1,000 km for a typical atmospheric neutrino with an energy of 1 GeV. *

The example we showed in Figure 16.4 assumed an equal mixture of the two masses. This actually fits the case of atmospheric neutrino oscillations very well. As shown in the figure, for a precise energy the neutrino oscillates, disappearing at certain distances and then completely reappearing at others. In the case of the atmospheric neutrino experiments, the data are an average over a set of energies, and so the oscillation is not seen. The result is that on average half of the v_μ disappeared.

What does this experiment tell us about the actual neutrino masses? The answer is that it only tells us about mass differences. More precisely,

---

* The use of electron-volts (eV) as a measure of mass was explained in connection with Table 6.1.

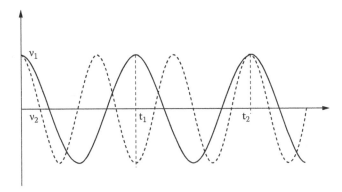

FIGURE 16.4   The curves show the phases of v_1 (solid) and v_2 (dashed) as a function of time (or of distance). They start out in phase; at time t_1 the phases are opposite and the sum of v_1 and v_2, which is v_μ, vanishes. The original neutrino is now in the state ($v_2 - v_1$), which is the v_τ. At a later time, t_2, v_1 and v_2 are back in phase, and the original neutrino is back to v_μ.

the oscillation length determines the difference in the squares of the masses.* Thus, the atmospheric neutrino results can be fitted with $m_2 = 0.05$ eV and $m_1 = 0$, or equally well with $m_2 = 0.13$ eV and $m_1 = 0.12$ eV.

More generally, we consider the case of unequal mixtures. Thus, v_e might be mainly v_1 and a little v_2, while v_μ would be mainly v_2 and a little v_1. The amount of mixing is measured by a mixing angle.

NOTE

1. No particle can move in vacuum faster than light's velocity in vacuum. However, light moves slower through water than it does through vacuum. And, it is possible for a particle to have such a high energy that it moves through water faster than light does. In such cases, light is emitted, known as Cerenkov radiation. This is the origin of the blue glow one sees in nuclear reactors.

---

* This holds in the good approximation that the values of mc^2 are much less than the total neutrino energy (see Appendix 5).

CHAPTER 17

Neutrinos from the Sun

A QUESTION THAT WORRIED Lord Kelvin in the nineteenth century was the origin of the sun's energy. Where did all that energy come from? If the sun were a large lump of coal, it would burn up very quickly. In fact, the best source of energy he knew of was gravitational energy. If the sun started very large and gradually collapsed on itself, the original potential energy would be converted into kinetic energy or heat. In this way, the sun could have been shining for only a few million years. That didn't seem long enough for the evolution required by Charles Darwin. We now know from radioactive dating that the solar system is 4.5 billion years old.

The answer came with the discovery of a new form of energy: nuclear energy. Ernest Rutherford proposed that the sun might be made of some radioactive material whose radiation provided the energy. An alternative proposed by two young physicists, Geoffrey Atkinson and Fritz Houtermans, in 1930 was that the energy came from nuclear reactions taking place in the hot center of the sun. When two nuclei are fused together to form a new nucleus, a million times more energy is generated than when carbon and oxygen are fused to form carbon dioxide. With this energy source the sun could shine for billions of years.

In the 1930s there was much progress in the study of nuclear reactions. A leader in this endeavor was Hans Bethe, who came as a refugee from Germany in 1933 to Cornell University, where he continued to do research for the next 65 years. In 1936 and 1937 he authored or coauthored a series of articles that occupied three whole issues of the *Reviews of Modern Physics* and provided a detailed account of nuclear theory. For years thereafter physicists prized these volumes and referred to them as the "Bethe Bible."

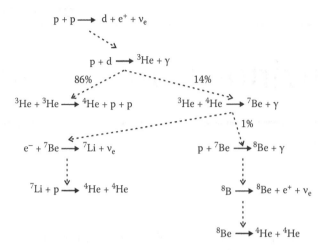

FIGURE 17.1   Chain of nuclear reactions inside the sun. There are three different chains or paths, each ending with the production of normal helium, ^{4}He (helium-4). The number associated with the symbol for an element is the mass number (number of protons plus number of neutrons) of the particular isotope of the element.

The sun, and indeed most of the universe, is made of hydrogen and helium. The sun must get its energy by a series of nuclear reactions that convert four protons into a helium nucleus. Because the helium nucleus contains two protons and two neutrons, two of the protons must be changed into neutrons in the course of the reactions. The set of the reactions that occur in the sun was proposed by Bethe together with Charles Critchfield in 1938 and is shown in Figure 17.1. Note that the first step is a weak interaction in which two protons fuse to form a deuteron, denoted by d, made of one proton and one neutron. This is the nucleus of the heavy hydrogen atom called deuterium. Because the interaction is weak, it proceeds relatively slowly, which helps to explain why the sun can keep shining so very long. For his discoveries concerning the energy production in stars, Hans Bethe was awarded the Nobel Prize in 1967.

How is it possible to check this theory concerning the energy production in the sun? How can you look into the center of the sun? The answer is that almost 3% of the energy produced comes in the form of neutrinos, and these can penetrate the entire sun with less than one in a thousand absorbed and reach the earth 8 minutes after they were produced. The heat and light we receive from the surface of the sun results from nuclear reactions that took place thousands of years earlier, since it takes that long for the energy to gradually be transferred from the center to the surface, but the neutrinos tell us about reactions that took place a mere 8 minutes earlier! More than a hundred billion neutrinos from the sun pass through your body every second, day and night; they interact so weakly that they do you no harm.

In the 1960s and for years thereafter John Bahcall carried out detailed calculations of the neutrinos to be expected from the reactions in the sun. It was Bahcall's calculations and his close interaction with experimentalists that played a major role in initiating the search for the solar neutrinos. One of us (Wolfenstein) enjoyed many interactions with Bahcall during his 37 years as professor at the Institute for Advanced Study in Princeton. Unfortunately, he died in 2005, before he had a chance at a Nobel Prize. As a first step in Bahcall's calculations, it was necessary to know the rate of the different nuclear reactions. This required experiments using beams from particle accelerators. In fact, it was the discovery in 1958 of the unexpected large rate of the fusion reaction between helium-3 and helium-4 to form beryllium-7 (^{7}Be) that led to the conclusion that solar neutrinos could be detected. The group at Cal Tech led by Willy Fowler (Nobel Prize, 1983) continued for many years to study nuclear reactions important for astrophysics. The initial reaction, the one that fuses the two protons into a deuteron, is too weak to duplicate in the laboratory, and so had to be calculated using weak interaction theory.

The rate of the fusion reactions depends very sensitively on the temperature, because the temperature determines the kinetic energy of the moving nuclei, and a little extra energy is very helpful in overcoming the electrical repulsion between two positive nuclei. Thus, it was necessary to determine the temperature near the center of the sun from the surface temperature, using a detailed analysis of the energy transfer. Given the rate of the individual reactions shown in Figure 17.1, it was then necessary to model the sequence of reactions, keeping track of where they occurred in the central region of the sun. The resulting spectrum of solar neutrinos is shown in Figure 17.2. There are three major groups of neutrinos: the highest energy from boron-8 are very rare, two out of every ten thousand neutrinos, but they are the easiest to detect; the second highest from beryllium-7 are 10% to 15%; and the very low energy neutrinos from the original pp reaction are the major component.

Given this wonderful possibility of looking into the sun, one might have expected there would have been many experiments studying these neutrinos. In fact, for 20 years, from 1965 to 1985, one man working almost

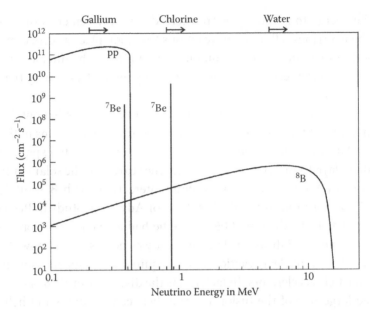

FIGURE 17.2 The calculated flux of solar neutrinos as a function of energy. Note that the scales are logarithmic. The ^{7}Be neutrinos are limited to two energies, while the others are distributed over a range of energies. The arrows on top show the thresholds for the different experiments.

alone detected the solar neutrinos. His name was Raymond Davis. In 2002 he shared the Nobel Prize in Physics, although in fact he was a chemist. As a radiochemist Davis had the only technique that was sensitive enough at that time, based on a proposal originally made by Bruno Pontecorvo in 1946, to detect neutrinos by a reaction in which a neutron in chlorine was changed to a proton, converting it to argon. A crucial step was a theoretical analysis by Bahcall that showed the probability of this reaction was twenty times larger than previously calculated.

The apparatus of Davis was a tank of 100,000 gallons of cleaning fluid. The neutrinos would convert a few of the chlorine atoms to argon, a noble gas, and the argon atoms would then be floating around in the cleaning fluid. After a month or so the argon atoms would be removed. The isotope of argon that is produced is actually radioactive, and so the number of argon atoms could be counted. The number of neutrinos would then be determined from the theoretical knowledge of the probability that a neutrino could change the chlorine into argon. This experiment was mainly sensitive to the very rare high-energy boron-8 neutrinos.

The prediction of Bahcall was that, on average, a little more than one argon atom a day would be produced in the 100,000 gallons. When Davis announced his results, he found a little less than half an atom a day. The neutrino flux was only a third of the prediction. Given the difficulty of the experiment and possible uncertainties in the theory, it took many years before most physicists were willing to believe there was a problem. Eventually, it became clear we could not avoid this solar neutrino problem. This serves as a wonderful example of problems in astrophysics: Was this something we didn't know about the sun or something we didn't know about the neutrino? Was this a problem for astronomy or a problem for particle physics?

The definitive answer came from the SNO experiment carried out in the depths of the Sudbury, Ontario, nickel mine. The experiment was originally proposed by Herb Chen in 1985, but he died before it could be carried out. The first results were reported in 2002. The basic idea was to use a water Cerenkov counter like the ones discussed above filled with heavy water instead of normal water. For heavy water, the hydrogen in the H_2O is replaced by deuterium; this means that the nucleus of the atom is a deuteron consisting of a proton together with a neutron. The experiment used 1,000 tons of heavy water; this was available in Canada because the Canadian government maintains a large supply of heavy water for use in its CANDU nuclear reactors. It was estimated to have a value of $30 million and was returned to the government when the experiment was completed. The location in the mine 6,800 ft below the surface provided a good shield against cosmic ray muons but, of course, the neutrinos had no trouble reaching the heavy water.

Two reactions are possible when the neutrino hits the neutron:

1. The neutron can change to a proton and an electron is emitted; this happens only for electron neutrinos.

2. The neutrino may scatter elastically off the neutron, giving the neutron a kick due to the neutral-current interaction; this happens with equal probability for all types of neutrinos.*

Like the Davis experiment, this experiment was only sensitive to the high-energy neutrinos. By counting the tracks of the electrons it was found that the flux of electron neutrinos was a third of the Bahcall prediction, in agreement with the Davis experiment. The second reaction could be measured by

---

* Only the neutron in deuterium is involved because the neutrons in the oxygen nucleus are too tightly bound.

detecting photons when the recoil neutrons were captured by the deuterion. By counting the number of photons, it was determined that the flux of neutrinos of all types was equal to the prediction. The conclusion was that the theory of the sun was correct and that two-thirds of the electron neutrinos produced in the sun arrived at the earth as the other types of neutrinos.

In order to really confirm the theory, it clearly was very desirable to observe the lower-energy neutrinos, which constitute more than 99% of all. The first such experiment was similar to that of Davis but used gallium in place of chlorine. The energy needed to transform the gallium into radioactive germanium was much less than that needed for detection in the chlorine experiment (see Figure 17.2); thus, all the different sources of neutrinos shown in Figure 17.2 could be detected. The expected signal would be mainly due to the lower-energy neutrinos, which constitute more than 99% of all the solar neutrinos. Although Davis made a proposal to carry out the gallium experiment, he could not get funding in the United States. Two gallium experiments were carried out: GALLEX in Italy and SAGE in the Soviet Union. As expected, they detected many more neutrinos than Davis, but the flux was only a little more than half of the Bahcall prediction. Presumably, almost half of the electron neutrinos had oscillated into the other types, but there was no way to detect them.

One question that was asked about the Davis experiment was how he could tell that the neutrinos came from the sun. It was in fact simply a process of elimination; there seemed to be no other source. The first experiment sensitive to the direction of the neutrinos was SuperKamiokande. Although this experiment could not detect the more probable reactions later detected by SNO, it could detect electron neutrinos by their elastic scattering from the electrons in the water. Because the electrons are so light, the effect of the scattering is to kick the electrons approximately in the direction the neutrinos were coming from. The resultant angular distribution is shown in Figure 17.3.

The possibility that solar neutrinos might be converted by neutrino oscillations was first proposed in 1969 by Gribov and Pontecorvo. They considered large oscillation lengths allowing neutrinos to oscillate as they traveled the 93 million miles to the earth. Such large oscillation lengths could never be detected using accelerator or reactors on earth. As the solar neutrino problem was studied, the possibility of shorter oscillation lengths was considered. In Japan a detector called KAMLAND was placed about 150 km from several powerful reactors; in 2002 a clear indication of oscillations was found from the distortion of the spectrum.

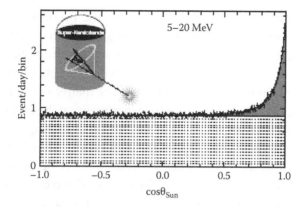

FIGURE 17.3 Angular distribution of electrons in SuperKamiokande. The direction of the sun corresponds to $\cos \theta_{Sun} = 1$. The flat part of the curve is a background not associated with solar neutrinos. The peak in the direction of the sun is due to the scattering of solar neutrinos. (Taken from SuperKamiokande's "Solar Neutrino Measurement in Super-Kamiokande" Pys. Rev D73, 11 2001[2006], pg. 12001–12021, fig. 40 [hep-ex/0508053v2]. With permission.)

The KAMLAND result makes clear that the oscillation length is much smaller than the radius of the sun. Neutrino oscillations could begin inside the sun as the neutrinos leave the central region and travel to the surface. As noted before, neutrinos traverse the whole sun as if the matter weren't there; so it was natural to believe that, as far as the neutrinos were concerned, traveling through the sun was essentially the same as traveling through empty space. Although neutrino oscillations were first considered in 1962, it was only 16 years later that one of us (Lincoln Wolfenstein) pointed out that oscillations in matter could be quite different from oscillations in vacuum. Glass and water may be transparent for light, but they do affect light because of the index of refraction. The sun may be transparent for the neutrino, but there is an index of refraction. The index of refraction is so close to 1 that it is of no importance until we consider oscillations.

Oscillations depend on considering neutrinos as waves, and the masses affect the wavelength. Similarly, the index of refraction affects the wavelength. What is important is the difference in the index of refraction between the electron neutrino and the others (muon neutrino and tau neutrino) because the electron neutrino interacts with electrons in atoms in a way that the others do not.*

---

* It is the neutrino electron elastic scattering from electrons due to W exchange that is different for electron neutrinos. As shown in optics, the index of refraction is related to the forward scattering amplitude.

Thus, the calculation of what happens to the neutrinos that start as electron neutrinos as they propagate through the sun depends both on the mass difference and mixing discussed above for the vacuum case and on the index of refraction difference. For the low-energy neutrinos, the vacuum effect dominates and the effect of the matter is relatively unimportant; thus, the result of the gallium experiment can be explained using the vacuum formula with large but not maximal mixing. The index of refraction is essential for the case of the high-energy neutrinos.

Consider the propagation of neutrinos from the center of the sun with just two types of neutrinos: there will be in general two propagating waves with different wavelengths. In the vacuum case the electron neutrino is a mixture of the two propagating waves, v_1 and v_2, as shown in Figure 16.4. For the case of the high-energy electron neutrinos, as they start out from the center of the sun, the mass difference is much less important in determining the wavelength than the index of refraction because the energy is so large compared to $m_2 - m_1$.* These neutrinos start out primarily as the wave v_S with the shorter wavelength because of the index of refraction with a very small probability of being v_L. As the neutrinos propagate, the wavelength of v_S becomes longer as the index of refraction decreases due to the decreasing density. The wavelengths of v_S and v_L as a function of distance from the center are shown in Figure 17.4. The neutrinos that

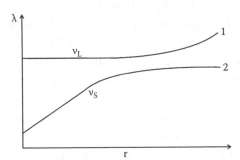

FIGURE 17.4 Qualitative diagram showing how the wavelengths of the neutrino waves propagating through the sun vary as a function of the distance, r, from the center of the sun for the high-energy neutrinos. The electron neutrinos produced in the sun start out in the state v_S and in the adiabatic approximation end up in this state, which is v_2, only 1/3 v_e. In this limit the state v_L is not occupied. In fact, there is a small probability that the neutrinos jump from the v_S state to the v_L when the wavelength difference becomes small. This probability is larger for the lower-energy neutrinos.

* What is important is the value of $m_2{}^2 - m_1{}^2$ divided by the energy, as discussed in Appendix 5.

start out in the ν_S wave will stay in this wave and not suddenly change wavelength to join the ν_L wave if the density change is sufficiently gradual. This is known as the adiabatic approximation and holds for the case of the high-energy neutrinos. As a result, the neutrinos emerge from the sun in the vacuum state with the shorter wavelength, corresponding to the state with mass m_2. The solar neutrino data is explained if the state ν_2 is 1/3 electron neutrino and 2/3 the other neutrino.

In this adiabatic approximation, there is no oscillation on the way between the sun and the earth because the neutrinos are not a mixture of two masses but have the mass m_2. However, when the neutrinos enter the earth there will be oscillations because the index of refraction due to earth matter plays a role. The calculation shows that there should be more solar neutrinos at night than during the day! The difference is too small to have been detected so far.

Although the importance of the index of refraction was first pointed out in 1978, it was only in 1985 that the correct application to the sun was made by two Russian physicists, Mikhaeyev and Smirnov. It is now known as the Mikhaeyev–Smirnov–Wolfenstein (MSW) effect. In discussing the transmission through the sun, they considered the neutrino as it passed through the sun as a mixture of ν_e and ν_μ (or ν_τ), starting as ν_e and then oscillating into ν_μ (or ν_τ) as the index of refraction decreased. However, in the high-energy limit it is simpler to understand this using the adiabatic approximation, as was made clear by a paper by Hans Bethe in 1986.

CHAPTER 18

Neutrino Astronomy

A New Way to Study the Universe

18.1 NEUTRINOS FROM STARS

The discovery of the solar neutrinos can be considered the beginning of a new type of astronomy. Neutrinos provide the possibility of exploring dense regions of astronomical objects from which no other radiation can escape.

There exists one other example of neutrino astronomy: the neutrinos from Supernova 1987a. On the morning of February 23, 1987, a young astronomer in Chile observed a bright new star in the sky: a supernova. Two water Cerenkov detectors were in operation at that time: Kamiokande in Japan and IMB in the United States. They carefully scanned their data over a period of days around that date. Eleven unusual events were detected in Kamiokande in a 10-second interval 3 hours before the supernova sighting. In the same 10-second interval, eight unusual events were detected in IMB. These events were interpreted as recoil electrons from neutrino interactions.

The supernova was identified as a type 2. This is believed to be a star that suddenly collapses when the nuclear fuel does not provide enough energy to overcome gravitational attraction. Within a fraction of a second, the center becomes extremely hot and dense. High-energy collisions in this dense plasma produce positrons and electrons, photons, and neutrinos. The only particles that could escape and release all this energy were the neutrinos. The nineteen neutrinos were a view into the center of the collapsing star.

The neutrinos marked the time of collapse. The light came later, when some of the energy was transferred to the outer surface of the star, a part that had not collapsed. The actual mechanism of this energy transfer is still not well understood. Although a supernova is spectacularly bright in emitted light (the peak luminosity of Supernova 1987a was 500 times that of the sun), the theory says that this is less than 1% of the energy of collapse; 99% comes out in neutrinos!

The location of this supernova was found to be outside the Milky Way in a neighboring galaxy called the Large Magellanic Cloud, 150,000 light-years away. The neutrinos had been traveling for 150,000 years and arrived just a few years after the detectors were ready! Given the large distance and the small size of the detectors, there must have been many neutrinos emitted. The calculation based on the nineteen neutrinos observed was that 30 billion trillion trillion trillion trillion neutrinos were emitted. This constitutes the first direct evidence of the gravitational collapse, and the observations were consistent with supernova theory.

What about the future? Taking into account oscillations, the solar neutrino data are consistent with the theoretical calculations for the sun. However, as of 2007 there was no direct observation of the main neutrinos, those from the original pp reaction (the top reaction in Figure 17.1) and those from the capture in beryllium. While we assume that the gallium experiment results were mainly due to these neutrinos, it is important to measure these two neutrino fluxes separately to quantitatively confirm the theory. An experiment called Borexino is operating in Italy designed to detect electrons scattered by the monoenergetic neutrinos that come from the electron capture in beryllium. The greatest challenge is to directly detect the main group of neutrinos, those that come from the very first reaction. This will be very difficult because of their very low energy; however, the detection of these neutrinos would provide proof to the world that we really do understand the major source of energy for mankind, the sun.

Sometime in the future, we expect a supernova in our own galaxy. Given the larger size of present detectors like SuperKamiokande and the closer distance, one might observe tens of thousands of neutrinos instead of only Nineteen. Information about the time dependence and the energy spectrum of these neutrinos would provide many constraints on the theory. Every large neutrino detector, no matter what its major purpose, should be part of a supernova watch. This means the detectors should never completely shut down, and that events should be timed accurately to a fraction of a second. It would even be possible to detect neutrinos from a supernova on the other side of our galaxy that could not be seen directly because the light was absorbed by the galactic center.

All types of neutrinos and antineutrinos are expected to be emitted from type 2 supernovas. They are produced by neutral current interactions in the very hot collapsed star. However, they are expected to emerge with somewhat different energies because of the interactions on the way out. The electron neutrinos, ν_e, are expected to have the lowest average energy because they have extra interactions with electrons, while the muon neutrinos, ν_μ, and tau neutrinos, ν_τ, would have the highest average energy. However, the observations on the earth will be affected by oscillations that take place from the time the neutrinos emerge from the dense center until they leave the star.

The neutrinos observed in 1987 were probably electron antineutrinos. This is because they can interact with the proton in hydrogen to produce a neutron and a positron, while the neutron in oxygen is too tightly bound to be excited by an electron neutrino. The electron neutrino could have been detected by elastic scattering from electrons, but the rate for this is much smaller. It will be desirable in the future to have different types of detectors that are sensitive to all types of neutrinos.

Over the last billion years, there have been a large number of supernova explosions. The neutrinos from these form a diffuse background of neutrinos everywhere, including here on earth. An ongoing effort in Japan is looking for those neutrinos using the SuperKamiokande detector; in particular, the goal is to look for energies larger than those of solar neutrinos. The observation of these neutrinos would allow us to look far into the past and obtain information on the rate of star formation and collapse over the last billion years.

Many sources of high-energy gamma rays have been observed in recent years. The theories that explain these make predictions as to the neutrino flux. Thus, the detection of very high energy neutrinos can help to determine the processes inside these sources. In order to observe these neutrinos, one must have an extremely large detector. The solution is to use a mass of polar ice or the water of the sea as the matter in which the neutrino interacts. The IceCube detector at the South Pole consists of 80 strings of photomultiplier tubes inserted in holes 2 km deep in the polar ice. It will detect neutrino interactions occurring within a kilometer cube of ice. The KM3NeT detector is located in the Mediterranean Sea near the coast of France and consists of similar strings stretching 250 m below the surface.

It may also be possible to learn something new about neutrinos from neutrino astronomy. If neutrinos from distant sources are due to pion decay, they should arrive with the ratio of v_e to v_μ to v_τ of 1:1:1. The decay chain of pion into muon plus v_μ, followed by muon decaying into electron plus v_e plus v_μ, yields one v_e and two v_μ. However, the atmospheric mixing with the large value of mass differences quickly mixes v_μ with v_τ, so that half of v_μ end up as v_τ.* Thus, the final state is an equal mixture of the three neutrino flavors. The relative phases between the three types of neutrinos are random, due to the complex way in which they are produced; this is referred to as an incoherent mixture. This can also be considered an incoherent mixture of the mass states, with the ratio of v_3 to v_2 to v_1 equal to 1:1:1. As a result, this ratio will not be changed by further oscillations.

If the ratio differs from this for neutrinos from a distant source, it is possible that some neutrinos have decayed during the long trip. For example, if all the v_3 decay but none of the v_2 and v_1 do, then the ratio of v_e to v_μ to v_τ would end up being 2:1:1. Thus, a neutrino decay time could be detected that is much longer than could be detected in any other way.

In 1950, the only astronomical observations involved visible light. The great advances of the second half of the twentieth century involved observations of radio waves, x-rays, and gamma rays—the whole electromagnetic spectrum. Neutrino astronomy is still in its infancy and can provide a new window on the universe in the twenty-first century.

18.2 NEUTRINOS FROM THE EARLY UNIVERSE: NEUTRINOS AS DARK MATTER

One of the great scientific questions is what the universe is made of. As astronomers have studied distant stars, they have found that the wavelengths of light emitted, what is called the spectrum, correspond to the same elements we know on earth. They consist of atoms made from protons, neutrons, and electrons. But is that all there is? The surprising answer found over the last 30 years is that most of the matter in the universe is another kind of matter that does not coalesce into shining stars: it is called dark matter.

The first strong evidence came from the study of spiral galaxies by the astronomer Vera Rubin. A spiral galaxy like our own Milky Way contains a very dense central region with stars rotating around it, something like our solar system with planets rotating around the sun. Rubin measured the velocity of stars at different distances from the center. In the case of planets, the velocity decreases with distance since the pull of the sun is less. Surprisingly, she found that the velocity stayed constant: the distant stars

---

* For simplicity we have not distinguished here neutrinos (v) from antineutrinos ($\bar{v}$).

were being attracted by a large mass distributed throughout the galaxy, not just the mass at the center. Various studies showed that this could not be a huge number of dark baby stars or interstellar gas. It was some new kind of weakly interacting matter that filled the galaxy, holding it together.

Many further observations have confirmed this conclusion. The gravitational mass that holds together galaxies and clusters of galaxies that form our universe consists of 15% ordinary matter and 85% dark matter. The nature of this dark matter is one of the major scientific questions facing us in the twenty-first century.

The study of the history, or evolution, of the universe is called cosmology. The starting point is the observation made by Edwin Hubble in the 1920s that distant stars were moving away from us with a speed that increased directly with the distance. The universe was expanding. This does not mean that we are in some special location at the center. An observer on a distant planet in another galaxy would make the same observation that the universe was expanding about him or her. It is as if we were molecules inside an expanding balloon and every molecule was receding from every other and the surface was unobservable.

If we assume the same laws of physics held in the past, we can look backward in time. Thus, at earlier and earlier times, the universe is compressed more and more and becomes very dense and very hot. Atoms cannot stick together any more, and we have a plasma of interacting electrons, protons, neutrons, positrons, photons, and neutrinos. This picture of the early universe was first promoted by George Gamow. Was there any way to test this picture, any fossil from the past? The answer was developed by Alpher and Hermann: as the universe expanded and cooled, the electrons would become fixed inside atoms and photons would not have enough energy to excite the atoms. So, they could not be absorbed and would be left over and fill the whole universe.

The wavelengths of these photons today would be around 1 cm and provide a cosmic microwave background (CMB). For many years there was no search for the CMB. Those who knew about possible detectors didn't take the theory seriously. Those who believed the theory thought the CMB was impossible to detect. In 1965, the CMB was detected by accident. Arno Penzias and Robert Williams had built a large detector for radio waves for applied work at Bell Labs and found some unwelcome noise at wavelengths around 1 cm; they published a paper describing this noise and in 1978 they won the Nobel Prize in Physics for it. They had discovered the CMB. The first fossil from the early universe had been discovered, and cosmology became an experimental science. In the last 15 years, there have been detailed observations of the CMB.

The same theory that predicted the CMB also tells us that there should be a background of neutrinos of all three types everywhere in the universe. These neutrinos are moving with such a low energy that it appears impossible to detect them. Now that we know neutrinos have a mass, it is clear that neutrinos are a part of the dark matter—the only part that we understand. What fraction of the dark matter is made up of massive neutrinos? A partial answer comes from detailed theories of the evolution of the universe.

After atoms formed in the early universe, they came together to form stars, and eventually stars gathered together in galaxies. Detailed computer calculations model this evolution. In this formation of structure, of galaxies and clusters of galaxies, the dark matter plays a crucial role since it is the main mass that pulls things together. The calculations show that if the dark matter consisted of very light particles, like neutrinos, the size of the structures in general would be larger than observed. Depending on the details of the calculations, it is concluded that at most 3% to 7% of the dark matter can be neutrinos. The rest should be much heavier particles, called cold dark matter, which remains a mystery.

Because the evolution of the universe determines the density of neutrinos, this limit provides a limit on the neutrino masses, which is the best limit today. With the addition of further data, this limit could be reduced. It has also been suggested that some details of galaxy structure might provide evidence for a nonzero mass close to the upper limit. The intersection of particle physics and cosmology represents an exciting frontier today.

The most common belief today is that the dark matter consists of some massive particle not yet discovered that interacts only weakly. Presumably this interaction produced it in the early universe, but some selection rule keeps it from decaying into the lighter particles of our standard model. However, it is possible that in places where the dark matter density is large enough, two dark matter particles may collide and annihilate by a weak interaction. A possible product of such annihilation would be a pair of neutrinos. Thus, one of the goals of high-energy neutrino telescopes like IceCube and KM3NeT is to look for neutrinos of a fixed energy coming from a source like the sun. The energy of the neutrino would be equal to the value of mc^2 for dark matter particles.

CHAPTER 19

Neutrino Mass and Physics beyond the Standard Model

THE DISCOVERY OF NEUTRINO oscillations told us that neutrinos do have mass. The explanation of the two different types of oscillations, the atmospheric and the solar, required two very different values for the mass differences. Therefore, there are three different masses. However, the oscillations only tell us about mass differences, allowing a whole set of possible masses. Some examples are shown in Table 19.1. The Mikhaeyev–Smirnov–Wolfenstein (MSW) solution to the solar neutrino problem requires m_2 to be greater than m_1; however, m_3 can be the largest or the smallest of the three masses. When the masses are relatively large, as in cases 4 and 8, the three masses are almost equal: they are referred to as quasi-degenerate.

The value of $m_2^2 - m_1^2$ is the same for all cases since it is determined from solar neutrino data and Kamland reactor data. The magnitude of $m_3^2 - m_2^2$, but not the sign, is the same for all cases; this has been determined from the atmospheric neutrino data and data from the MINOS experiment at Fermilab on the disappearance of muon neutrinos beamed to a detector in Minnesota. In the first four cases m_3 is the largest; this is called the normal hierarchy. In the last m_3 is the smallest; this is called the inverse hierarchy. For each case, the range of values for M_{ee} is given; this

TABLE 19.1   Sets of Values of Neutrino Masses That Fit Present Data

m_1	m_2	m_3	M_{ee}
0	0.009	0.05	0.003
0.02	0.022	0.055	0.007–0.021
0.051	0.052	0.073	0.017–0.051
0.1000	0.1004	0.112	0.033–0.100
0.050	0.051	0	0.017–0.051
0.053	0.054	0.02	0.018–0.053
0.0728	0.0710	0.05	0.024–0.071
0.1125	0.1120	0.10	0.037–0.112

Note: All masses are in electron volt (eV). The column M_{ee} indicates the range of values of the electron neutrino mass that determines the rate of neutrinoless double beta decay. Future experiments using 1 ton of material might detect neutrinoless double beta decay if $M_{ee} > 0.05$ eV.

is the quantity that determines the rate of neutrinoless double beta decay, to be discussed below.

One way to determine the mass of the electron neutrino was pointed out in Fermi's original paper. In nuclear beta decay, the maximum energy of the emitted electron decreases as the mass of the neutrino increases, because some of the available energy must be used to provide the mc^2 of the neutrino. Thus, by studying the top end of the electron spectrum it is possible to put limits on the neutrino mass.* The most accurate results come from the study of tritium decay (tritium is an isotope of hydrogen with one proton and two neutrons that decays into an isotope of helium, helium-3). The results from a group in Mainz, Germany, give an upper limit of about 2 eV. While the electron neutrino is a mixture of two different masses, the difference between the two is so small that this limit clearly applies to both. A new experiment labeled Katrina is planned with a hope of bringing the limit down by a factor of 10.

If neutrinos were really massless, they would travel with the velocity of light. Neutrinos with mass have a slightly lower velocity, which depends on the energy; the higher the energy, the closer to the velocity of light. The neutrinos from Supernova 1987a had a variety of energies, but all arrived within a few seconds of each other. If the neutrino mass were large enough, the low-energy neutrinos would have come later since they were moving

---

* It is actually the shape of the spectrum at the highest energy that is used, as seen in Figure 5.3.

more slowly. Given the limited statistics, the best limit from this observation was about 20 eV.

By far the best limit on neutrino masses today comes from the cosmological analysis discussed in the last section. Because this is an indirect determination depending on cosmology theory, it would be very good to find a more direct verification of this limit, but that does not seem very likely. The cosmological limit is on the sum of the masses of all three neutrinos since all types of neutrinos are expected in the background left over from the early universe.*

Depending on the details of the analysis, the upper limit on the sum of the masses is given as between 0.3 and 0.7 eV. It is expected that these limits will be improved as more observations and analyses are made. If one accepts the limit of 0.3 eV, then one sees from Table 19.1 that at least one neutrino must have a mass between 0.05 and 0.1 eV.

The standard model starts out with a gauge theory in which all the particles are massless: electrons, neutrinos, and so forth. There are two types of massless electrons to start with: the left-handed electron, e_L, and the right-handed electron, e_R. As required by the general CPT symmetry, e_L† has an antiparticle that is a right-handed positron. Similarly, e_R has an antiparticle that is a left-handed positron. When the gauge symmetry is broken by the Higgs mechanism, the e_L and e_R are combined into a normal massive spin 1/2 particle known as a Dirac particle. The electrons have two spin states and the positrons have two spin states.

In the standard model, there is no right-handed neutrino, and so neutrinos remain massless. One reason for leaving it out is that a right-handed neutrino being electrically neutral would have no standard model interactions. Another reason, of course, was that it was believed that neutrinos were massless. A minor change in the standard model would be to add a right-handed neutrino and give neutrinos a mass just like the electron. This would then be a Dirac particle or a Dirac neutrino.

There is, however, a problem. The heaviest of the neutrinos is several million times lighter than the electron. If the neutrino gets its mass in the same way as the electron, why should there be such a great difference in their mass? At this point, we have to face one of the great problems of our standard model. It was invented because of the beauty of the idea of a gauge

---

* All types of neutrinos can be produced by neutral current interactions of electrons with positrons.
† See discussion of C, P, and T in Chapter 7.

theory, and it has been validated by a succession of experiments. However, it contains a set of arbitrary numbers, most notoriously the masses of the elementary particles: the mass of the top quark is three hundred thousand times that of the electron, and that is just put in by hand with no explanation. Clearly the neutrino mass could similarly be put in by hand, but the very small masses of all three neutrinos seem to suggest that something beyond the standard model is in play. There is a general belief that at some higher energy scale beyond that of the W and Z masses there is some new physics, and that it might show up as small changes in the standard model predictions. Could neutrino masses be such a signal?

If we start with a left-handed neutrino and it has a mass, then there must be a right-handed neutrino. Because it has a mass, the neutrino moves with less than the velocity of light. Thus, it is conceivable that you can run past it so fast that its apparent direction of motion is reversed but its spin direction is unchanged, and so it is right-handed. However, it is possible to avoid having a Dirac neutrino. Because it has no charge, the neutrino could be its own antiparticle. The right-handed neutrino would be identical to the right-handed antineutrino that we know must exist. A neutrino that is its own antiparticle is called a Majorana neutrino since the concept originated with a young Italian physicist, Ettore Majorana.

Majorana was a precocious genius who was held in the highest regard by Enrico Fermi, who hired him for his Rome physics institute. He was also a deeply disturbed person. He was as fast in abstract thought as he was inept at social interaction. He disappeared from a ship at the age of 31; most likely it was a suicide.

In order for neutrinos to be massive Majorana particles, some fundamental new physics beyond the standard model is needed. In the standard model, there is a conservation law of lepton number. By assigning lepton number –1 to charged leptons and to neutrinos and +1 to their antiparticles, in a weak interaction the creation of a particle with lepton number –1 is always accompanied by a particle with number +1. However, if massive neutrinos are Majorana particles, we can no longer have this conservation law. Indeed, the interaction that gives neutrinos their mass mixes neutrinos with antineutrinos and violates lepton number by 2.

How can we test this possibility? We could look for neutrino reactions that appeared to violate the lepton number by two units. The problem is that in the limit that the mass is zero, such processes disappear, and so their rate must be proportional to some power of the mass. Since the masses are very small, such processes seem impossible to detect. The best hope for detecting the violation of a lepton number comes from neutrino-less double beta decay. There exist certain radioactive isotopes of nuclear charge Z that can decay to a nucleus of charge Z + 2 emitting two elec-trons and two antineutrinos, as illustrated in Figure 19.1. However, if the lepton number is violated, the decay could take place with the emission of two electrons and no antineutrinos. This is illustrated by the Feynman diagram in Figure 19.2. The rate is proportional to the square of the mass of the electron neutrino, M_{ee}. Of course, we have said the electron neu-trino doesn't have a definite mass since it is in general a mixture of the three mass states. In fact, it turns out that it is a mixture mainly of two states, v_1 and v_2, with little or no v_3. It is approximately 2/3 v_1 and 1/3 v_2 as required to fit the solar neutrino data. Thus, one guesses that the rate will depend on (2/3 m_1 + 1/3 m_2). This could be the answer, but it turns out that the + sign could be a minus sign, so that the two masses cancel instead of adding. In general the mass of the electron neutrino that enters can be

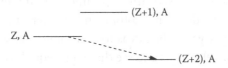

FIGURE 19.1 Nuclear energy levels allowing double beta decay. The three levels have the same mass number A (equal to the number of protons plus neutrons) but differ in Z, the number of protons. The only possible decay for (Z,A) is to emit two electrons going to (Z + 2,A) because the intermediate state has too high an energy.

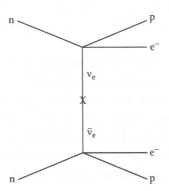

FIGURE 19.2 Feynman diagram for double beta decay in which two neutrons in a nucleus change into two protons and two electrons. The x stands for the effect of a Majorana neutrino mass, which combines a neutrino with an antineutrino.

TABLE 19.2   Some Proposed Experiments Looking for Neutrinoless Double Beta Decay and Their Estimated Sensitivity to the Electron Mass, M_{ee}

Experiment	Isotope	Mass (tons)	M_{ee} (eV)
EXO	^{136}Xe	1	0.05
MOON	^{100}Mo	10	0.03
CUORE	^{130}Te	0.75	0.1
MAJORANA	^{76}Ge	0.5	0.05

Note: The mass is the total in tons of the sample, which is not entirely the isotope of interest.

anything between those two extremes. The possible range of values for the mass that determines the rate of neutrinoless double beta decay is shown for different cases in Table 19.1.

Experimental searches for neutrinoless double beta decay carried out so far have not had the sensitivity to explore such low values of M_{ee}. In order to reach these values, future experiments will require the use of the order of a ton of the decaying isotope and the detection of as few as one hundred decays per year. The key is to accurately measure the sum of the energies of the two electrons and see that it just equals the available energy in the transition shown in Figure 19.1.* It is also necessary to minimize possible backgrounds, which requires placing the experiments in underground laboratories to provide shielding from cosmic rays. A number of experiments have been carried out, and new ones are planned at the large underground laboratory in the Gran Sasso in the Appenine Mountains in Italy. Other experiments will be located in the Sudbury mine in Canada and in an expanded underground laboratory in the Homestake mine in the United States, the place where Ray Davis did his pioneering solar neutrino experiment.

There are a large number of proposals for such experiments, but it will take a number of years before any of them can be carried out. Four of these proposals are listed in Table 19.2 with their approximate sensitivity to M_{ee}. In fact, this value is very uncertain because it depends on complicated nuclear physics calculations. Thus, it will be very useful to perform experiments with different isotopes with the hope that at least one will provide definitive evidence for neutrinoless double beta decay, and thus show that

---

* The decay can also take place with the emission of two electrons and two antineutrinos, which does not require any new physics. In this case, the two electrons are less energetic since the antineutrinos take away some of the energy.

the lepton number is violated. However, it is possible that M_{ee} is so small that none of these ambitious experiments will be sensitive enough.

Even if neutrinos are not Dirac particles, there is the possibility that right-handed neutrinos exist that are not the antineutrinos discussed above. These would also be Majorana particles and obtain their mass by some new physics. There are many possibilities for the values of the masses of these neutrinos, but the usual idea is that they might be much heavier than the neutrinos we know. Becase the weak interactions only affect the left-handed neutrinos (and right-handed antineutrinos), these neutrinos have none of the standard model interactions and so are called sterile neutrinos and are sometimes denoted by ν_S. In contrast, we will call the usual neutrinos active neutrinos.

There is one interaction consistent with the standard model that the sterile neutrinos could have; that is the interaction between a sterile neutrino, a left-handed neutrino, and the Higgs boson. When the Higgs boson gets a vacuum expectation value,* this can cause a small mixing between the sterile neutrinos and the active neutrinos.†

An experiment called LSND at Los Alamos seemed to give evidence for a small oscillation of a muon neutrino into an electron neutrino, requiring a mass difference greater than that for the two oscillations we already knew. It seemed possible to fit these data by adding one or more sterile neutrinos to the mixture with mass around 1 eV, a factor of at least ten larger than any of the active neutrinos. The results of this experiment have not been confirmed.

More generally, it is interesting to analyze every neutrino oscillation experiment, including the possibility of oscillation into sterile neutrinos. In the case of the atmospheric neutrinos the main observation was the disappearance of muon neutrinos; because no excess of electron neutrinos appeared, we assumed that the oscillation was into tau neutrinos, but it could have gone to sterile neutrinos. Although it is difficult to directly detect tau neutrinos, there is a way to distinguish the oscillation to tau neutrinos from the oscillation to sterile neutrinos. The oscillations take place in matter (in fact, they go through the whole earth); for oscillations

---

* See Chapter 8 and Part D for explanations of the vacuum expectation value.
† There is the possibility that the sterile neutrinos get a large Majorana mass by some new physics, but the active neutrinos do not. In this case, the active neutrinos will pick up a small fraction of the mass because of the mixing. This is called the seesaw theory because the larger the mass of the sterile neutrino, the smaller the fraction of that mass the active neutrinos acquire.

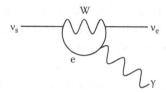

FIGURE 19.3 Decay of a sterile neutrino (v_S), producing an electron neutrino (v_e) and a photon (γ). Since the v_S mixes a little with v_e, it has the small coupling to e and W.

from muon neutrinos into tau neutrinos, the earth plays no role because both neutrino species have the same interaction with matter. However, while sterile neutrinos have no weak interactions, the tau neutrinos have the neutral current interaction. As a result, the dependence of the wavelength of the oscillation on energy is different for the two cases. A study of the energy dependence of the oscillation has shown that most of the oscillation does not go to sterile neutrinos.

Another interesting possibility is a sterile neutrino with a mass around 5 keV—thousands of times heavier than the active neutrinos, but still a hundred times lighter than the electron. It is assumed that these neutrinos have a very small mixing with the active neutrinos, so that they cannot affect oscillation experiments, but the mixing is sufficient to create them in the early days of the universe. They now could be the dark matter that pervades the universe. As a result of this mixing, they could decay into an electron neutrino and a photon, as shown in Figure 19.3. Since the dark matter is everywhere around us, the earth would be bombarded by these photons. Astronomers with x-ray telescopes have searched for an x-ray of a fixed energy around 5 keV and so far have found none.

CHAPTER 20

CP Violation in Neutrino Mixing?

20.1 EXPERIMENTAL SEARCH

Neutrino oscillations occur, as noted, because the states with definite weak interaction properties (v_e, v_μ, and v_τ) are mixtures of states with definite masses (v_1 with mass m_1, v_2 with mass m_2, and v_3 with mass m_3). This is completely analogous to the quark mixing described by the Cabibbo–Kobayashi–Maskawa (CKM) matrix. That naturally raises the question of whether neutrino mixing also leads to CP violation. The observation of another source of CP violation would be very interesting, particularly if we believe that neutrino masses are telling us something about physics beyond the standard model.

The neutrino mixing matrix analogous to the quark CKM matrix is called the PMNS matrix after Bruno Pontecorvo, who first suggested mixing, and Maki, Nakagawa, and Sakata, who first wrote down the matrix for two types of neutrinos. On the basis of present data, it can be written to a good approximation as in Table 20.1. The entries for v_3 are determined from the atmospheric oscillation, and those for v_2 from the solar neutrinos. However, one mixing angle, labeled θ_{13} (pronounced "theta-1-3"), is not known and could be zero. It measures the mixing of the muon neutrino, v_μ, with the electron neutrino, v_e, in the atmospheric oscillation. As noted in discussing Figure 16.3, the atmospheric data give no indication of an oscillation to v_e.

TABLE 20.1   The PMNS Neutrino Mixing Matrix

	ν_e	ν_μ	ν_τ
ν_3	θ_{13}	$\dfrac{1}{\sqrt{2}}$	$-\dfrac{1}{\sqrt{2}}$
ν_2	$\dfrac{1}{\sqrt{3}}$	$\dfrac{1}{\sqrt{3}}$	$\dfrac{1}{\sqrt{3}}$
ν_1	$-\sqrt{\dfrac{2}{3}}$	$\dfrac{1}{\sqrt{6}}$	$\dfrac{1}{\sqrt{6}}$

The best limit on θ_{13} comes from nuclear reactor data. Here one looks for the disappearance of electron antineutrinos in a detector at some distance from the reactor. The failure to see any disappearance puts a limit on the possibility that the ν_e have oscillated into ν_μ or ν_τ. For a typical reactor neutrino, the oscillation length associated with ν_3 is about 1 to 2 km. The CHOOZ reactor in France gave a limit on θ^2_{13} of about 0.04. New reactor experiments plan to increase the sensitivity by a factor of 10 and may be able to determine a nonzero value, if it is not too small.* One of these is a joint U.S.-Chinese collaboration at the Daya Bay Nuclear Power Facility, 55 km northeast of Hong Kong. A 20-ton liquid scintillator detector will be located underground in the nearby mountains, 2 km away from three powerful nuclear reactors. There will also be two detectors close to the reactors, so that it is possible to carefully compare the distant neutrino flux to the near one, to detect a possible oscillation. The experiment should start by 2011.

Another type of search for θ_{13} involves using muon neutrino beams from an accelerator and looking for the appearance of electron neutrinos. This is being carried out using a new accelerator in Japan sending a beam to the SuperKamiokande detector 300 miles away. (Because the energy of the beam is 200 times as great as that of reactor neutrinos, the oscillation length is 200 times longer.) A similar experiment called NOVA is being planned; it will send a beam from Fermilab near Chicago to a new detector in Minnesota.

* There is an oscillation of ν_e to ν_μ, as shown by the solar neutrinos as well as the KAMLAND reactor experiment, but this has a much larger oscillation length. Thus, this oscillation cannot be detected by detectors at a distance of only 1 or 2 km.

As noted for the case of the CKM matrix, there is one phase that can be added to the matrix that results in CP violation.* However, if one of the elements of the matrix is zero, this phase can be eliminated. For practical purposes we can attach the phase to θ_{13}. Therefore, CP violation in neutrino oscillations requires a nonzero value for θ_{13}, which is one reason the search for θ_{13} is so important.

If θ_{13} is not too small, it is possible to do experiments to demonstrate this CP violation. A direct search might seem to involve comparing the oscillation of muon neutrinos to electron neutrinos with the oscillation of muon antineutrinos to electron antineutrinos, which could be done using beams from an accelerator. However, such an experiment can only be done sending beams through the earth, and as discussed before, oscillations involving electron neutrinos are affected by the material of the earth. Thus, to truly directly test for CP violation, if muon neutrinos traveled through the earth, the muon antineutrino beam would have to travel through the antiearth!

However, with enough experiments it would be possible to take into account the effect of the earth and use such experiments to determine the CP-violating phase if it is large enough. For example, it is proposed to send a beam of muon neutrinos with a broad range of energies from Fermilab to a large detector in an underground laboratory in South Dakota 1,280 km away. The appearance of electron neutrinos as a function of the neutrino energy for such an experiment is illustrated in Figure 20.1.[1] It is seen at the lower energies that the result has a large dependence on the phase δ of θ_{13}. A variety of experiments like this could demonstrate a nonzero value of the CP-violating phase δ, if θ_{13} is large enough.

The reason for the dependence on δ is that both the shorter wavelength oscillation (the atmospheric) and the longer wavelength (the solar) contribute here. The addition of the effects of the two oscillations depends on the relative phase δ. Thus, there is sensitivity to the phase, even though you are not observing CP violation.

20.2 LEPTOGENESIS

One of the reasons for the interest in CP violation is the suggestion of Sakharov that this might explain the asymmetry between matter and antimatter in the universe. However, the CP violation of the standard model is

---

* In the case of Majorana neutrinos there is the possibility of two additional phases, but these do not affect oscillations and are probably undetectable.

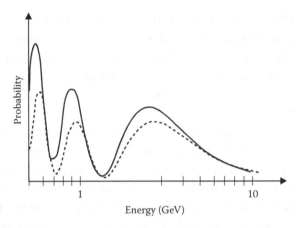

FIGURE 20.1 Probability for the appearance of an electron neutrino from a muon neutrino with CP-violating phase $\delta = 45°$ (solid line) and $\delta = 0°$ (dotted line). Here $\theta_{13} = 0.1$.

not sufficient to explain this. This raises interest in the possibility that CP violation in the neutrino sector may help to provide an answer.

Processes observed in experiments all consistent with the standard model display two conserved quantum numbers called baryon number, B, and lepton number, L. The proton and neutron have B = 1, and their anti-particles have B = −1. The u and d quarks have B = 1/3 and their antiparticles B = −1/3. Similarly, the electron and its neutrino have L = −1, and their anti-particles have L = 1. Thus, in beta decay, neutron decays into proton, electron, and electron antineutrino, and we have B and L separately conserved.

However, there is no fundamental symmetry in the standard model that guarantees B and L conservation. It turns out there is a fundamental conservation of B-L and it is possible to have processes that convert the lepton number into a baryon number conserving B-L. Such processes are essentially impossible at low energies but can take place at very high energies, and so could be significant in the history of the universe.

For Majorana neutrinos, neutrino mass is associated with the violation of lepton number. As discussed above, there is the possibility that neutrino mixing might violate CP invariance. Combining these two symmetry violations, there could be processes in the early universe that led to a nonzero value of L, and some of this lepton number could have been converted to a baryon number, B, thus explaining the matter-antimatter asymmetry of the universe. This is the concept called leptogenesis originally suggested by two Japanese physicists, Fukugita and Yanagida.

However, if we have only the three known neutrinos, this idea does not work, even if the CP violation turns out to be large. This brings us again to the sterile neutrinos. Suppose one or more of these has a very large mass and they are produced in the early universe. They could then decay to a light neutrino and a Higgs boson with the decay violating the lepton number. Considering a single sterile neutrino, N, it could decay as

$$N \rightarrow nu + X \text{ (Decay a)}$$

$$N \rightarrow nubar + X \text{ (Decay b)}$$

where X could be the Higgs boson and nu and nubar could be mixtures of the three types of active neutrinos. As a result of CP violation, decay b may be greater than decay a so that, when all N decay, there will be an excess of nubar over nu, and thus a net lepton number L. If this happens early enough in the early universe, so that all particles have a high energy, this L can be converted to a baryon number, B. This is the present proposal for leptogenesis.

If double beta decay is discovered showing L is not conserved, and if long baseline experiments show that CP violation is present in the neutrino mixing matrix, then two fundamental ingredients for leptogenesis will have been uncovered. However, direct detection of the sterile neutrinos N that start it all is presumed to be impossible because they are expected to be much too heavy to produce, and all of them will have decayed early in the history of the universe.

NOTE

1. Sketch based on formulas presented in the article "Intense Neutrino Beams and Leptonic CP Violation" by William Marciano and Zohreh Parsa, contributed to the 22nd International Conference on Neutrino Physics and Astrophysics (Neutrino 2006), Santa Fe, New Mexico, June 13–19, 2006. e-Print: hep-ph/0610258.

PART D

The Hunt for the Mysterious Higgs

Of all the particles of the standard model, the most mysterious is the Higgs boson. It is the one particle of the model for which there is no direct observation.

The standard model starts as a gauge theory that requires massless fermions and gauge bosons. The Higgs interacts with all the fermions with very different interaction strengths. Then we are told in the usual presentation of the theory that the Higgs field acquires a vacuum expectation value (vev) so that the interaction gives the fermions their very different masses. It also gives masses to the gauge bosons as a result of the gauge interaction of the Higgs.

Taken literally, the vev means that everywhere in space there is a constant Higgs field. The energy density of this field would be absurdly large and completely dominate the universe, altering cosmology as we know it. Thus, the vev makes the story of the Higgs boson even more mysterious.

It is important to note that the vev of the Higgs field plays no role in calculations of physics processes once the masses of the particles are specified. Why not just add mass terms to the original gauge theory and forget about the Higgs? The problem is that if we decided to arbitrarily break the gauge symmetry by giving masses to the W and Z gauge bosons, then it would be impossible to make any accurate calculations of weak interactions. What is important is the Higgs particle;* detailed calculations of weak interaction processes require the inclusion of virtual Higgs particles in order to give accurate answers. In fact, without the Higgs particle in such a theory you get crazy answers, with reaction rates equal to infinity. Thus, the Higgs interactions are essential. Indeed, the very precise data on various weak interactions discussed in Chapter 22 constrain the mass of the Higgs to lie between 115 and 160 GeV with 95% confidence.

More generally, we should say that there must be something that plays the role of the Higgs, but that it may not be the single boson of the standard model. In some extensions of the standard model, there are two or more bosons that share the role. In other theories there are new types of quarks, and the role of the Higgs is played by quark-antiquark pairs. Indeed, one can say that the Higgs is still a total mystery, and a major goal of the new Large Hadron Collider (LHC) accelerator is to solve this mystery.

---

* The Higgs particle, like all other particles, is described by a field. But it is really the value of the field above the constant value (the vev) that is relevant.

CHAPTER 21

Why We Believe
in the Higgs

21.1 THE STANDARD MODEL NEEDS
THE HIGGS MECHANISM

We saw that the Higgs mechanism is an essential part of the standard model. This mechanism is needed because of a contradiction between two features: the initial theoretical desire to have a gauge theory, leading to massless fermions and gauge bosons, on the one hand; and the well-known experimental fact that electrons and other fermions have mass, the same holding for the weak gauge bosons, on the other hand. The Higgs is related to the problem of mass; it was invented in order to give mass to both the gauge bosons and the fermions.

The masses of the quarks are related to a set of 18 complex parameters (the so-called Yukawa couplings linking the quarks with the Higgs field). These parameters can accommodate the known masses of the quarks, but they do not explain why these masses are what they are. Why is the top quark one hundred thousand times heavier than the up quark? We do not know. We can fit them in the theory, but the theory does not predict the masses of the quarks. This is known as the flavor problem. The same flavor problem exists for leptons. So, the Higgs mechanism is related to the flavor problem.

Similarly, we can fit the four parameters of the Cabibbo–Kobayashi–Maskawa matrix, but we cannot predict them. Of these four parameters, the three mixing angles are usually subsumed with the quark masses under the flavor problem. But the fourth parameter, the complex phase in the Cabibbo–Kobayashi–Maskawa matrix, is related to CP violation. Thus, the Higgs mechanism is also related to CP violation.

The energy scale of Higgs-related phenomena is set by the 250 GeV of its vacuum expectation value. The fact that only the top quark mass is of this scale, all other fermions being much lighter, poses the flavor problem mentioned before. But there is another problem. Gravity becomes relevant at the enormous energy scale of 10^{19} GeV, known as the Planck scale. Why is the weak (Higgs) scale so much smaller than the Planck scale? We do not know. This is known as the hierarchy problem. It is difficult to justify this energy scale difference in any natural fashion that does not go way beyond the standard model.

Thus, it appears that all the problems with the standard model, the features that seem most arbitrary, are related to the Higgs mechanism. We conclude that this mysterious Higgs mechanism is an essential feature of the standard model and is closely related to all the other mysteries concerning the standard model. It is befitting that the Higgs particle, the material side of the Higgs mechanism that holds such a central role in the standard model, should be the last to be accessed experimentally.

21.2 THEORETICAL CALCULATIONS NEED THE HIGGS PARTICLE

Imagine that we did not know about the Higgs mechanism. We knew about the $SU(2)_L \times U(1)$ gauge symmetry, but we did not know how to give masses to the particles. Could we still be led into the Higgs boson for reasons other than the need to generate the masses of all particles? The remarkable answer is yes!

To show how this was done, we consider the example of the scattering of two W bosons from each other. While this is not something that has been measured experimentally, it should be possible to calculate theoretically. In particular, it turned out to be theoretically interesting to consider the case when both W bosons were polarized longitudinally; that is, the spins were pointed along the direction of motion. This was calculated using the diagrams in Figure 21.1.* When this calculation was carried out, it gave crazy answers. As the energy of the WW collision is increased, the probability of the reaction increases without bound. This is clearly impossible

---

* Recall that the SU(2) gauge theory requires the existence of WWZ and WWWW vertices, and predicts their strength.

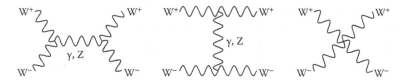

FIGURE 21.1   Tree-level diagrams involving only the gauge bosons that contribute to $W^+ W^- \to W^+ W^-$ scattering with longitudinal polarizations.

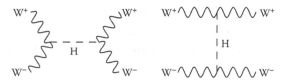

FIGURE 21.2   Tree-level diagrams involving the scalar particle that contribute to $W^+ W^- \to W^+ W^-$ scattering with longitudinal polarizations.

since probabilities can never be larger than 1. This sort of argument is known as a unitarity consideration; we say that the result is inconsistent with unitarity. We conclude that something else is needed, if we are to make sense of WW scattering.

We can save this process by inventing a scalar particle (we will subtly call it H) that couples to a W^+ and a W^- with some strength λ. This will lead to the two new diagrams in Figure 21.2. The diagrams in Figure 21.2 solve the problem with the diagrams in Figure 21.1, as long as the coupling λ equals the product of the weak coupling constant, g, and the mass of the W bosons. Thus, a massive spin 1 gauge boson begs a spin 0 particle to cure the infinities, with a coupling proportional to the gauge boson mass. A similar reasoning applied to $e^+ e^- \to W^+ W^-$ with longitudinal polarizations shows that the scalar must also couple to the electrons, proportionally to the electron mass.

The conclusion is that even if the Higgs mechanism is not the final answer, something must be faking the Higgs particle with great precision. This will be even clearer in Chapter 22, in relation to the precision electroweak measurements performed over the last 20 years at CERN, SLAC, Fermilab, and elsewhere.

Having established that a Higgs particle is needed, and that it couples to gauge bosons proportionally to their masses, we can calculate the full $W^+ W^- \to W^+ W^-$ scattering using Figures 21.1 and 21.2. With the Higgs contributions in Figure 21.2, the result depends on the Higgs mass and no longer grows alarmingly with energy.

Calculations like these were carried out before there were any relevant experiments. It was concluded that, in order to obtain reasonable results, there were limits on the possible Higgs mass. An early result by Lee, Quigg, and Tacker in 1977 proved that the Higgs mass should be less than 1 TeV (pronounced "tera-electron-volt" and corresponding to 10^{12} eV = 1,000 GeV); analysis of experiments discussed in the next chapter now suggests the mass is less than 160 GeV. It was concluded that the Higgs particle, or whatever mechanism takes its place, should occur at an energy close to the one explored by LHC. This explains the excitement behind the LHC: it will peek into the only remaining puzzle of the standard model, the mechanism of electroweak symmetry breaking.

CHAPTER 22

What We Know from Experiment So Far

22.1 INDIRECT EXPERIMENTAL EVIDENCE FOR THE HIGGS BOSON: THE IMPORTANCE OF VIRTUAL EFFECTS

In the standard model the weak and electromagnetic interactions are determined by three parameters: the W interaction strength (g), the B interaction strength (g′, pronounced "g-prime"), and the vev of the Higgs (v). In place of g′, it is customary to introduce θ_W (pronounced "theta-W"), known as the Weinberg angle.* As described in Figure 8.2, one combination of B and W^3 forms the photon, and the other combination forms the Z boson that mediates neutral current interactions. The mass of the W^+ or W^- is given by g × v^2. The strength of the original weak interaction given by Fermi's constant is determined by g and the mass of the W. Thus, knowing Fermi's constant and the mass of the W determines g and v. The strength of the electromagnetic interaction given by the charge of the electron, e, is given by a combination of g and g′. Thus, knowing g, the value of e determines g′ or θ_W.

These three parameters also determine the mass of the Z boson and the strength of the neutral current interactions. The great success of the standard model lies in the fact that all weak interaction data can be fit with these parameters. However, as the data became more precise, this required considering more complicated Feynman diagrams than those of Figure 8.1b or Figure 8.3. It required including the effects of virtual

---

* $\sin^2\theta_W = g'^2/(g^2+g'^2)$.

particles other than the W or Z. These are particles that are not in the initial or final state, but the influence of which must be included in a precise calculation. We have seen an example of virtual particles in quantum electrodynamics, where virtual photons can affect the mass of the electron, as seen in Figure 22.1. In the same way, a particle whose mass is too large to be produced directly by current facilities may be felt indirectly through its effects as a virtual particle in some loop.

A very interesting historical example concerns the top quark. Indeed, before it was detected directly, a good bound was already known on the top quark mass, because the top quark affects the masses of the W and Z bosons. At the lowest order in perturbation theory (so-called tree level) the mass of the W boson equals the mass of the Z boson times the cosine of the Weinberg angle, θ_W. But, loops with virtual quarks destroy this equality. Such violations are measured by a quantity known as $\Delta\rho$ (pronounced "delta rho"), which is known experimentally with great precision.

Were it not for loop effects, $\Delta\rho$ would be zero. Reversing the argument, measurements of $\Delta\rho$ can be used to probe virtual particles arising in loops. In the standard model, the most important contributions to $\Delta\rho$ come from the diagrams in Figure 22.2. When the calculation corresponding to the diagram in Figure 22.2a is made, the result grows with the square of the top quark mass; it is very sensitive to its value. As a result, even before it was measured directly, the top quark mass was known from the combination of various observables, including $\Delta\rho$, to lie between 150 and 210 GeV.[1]

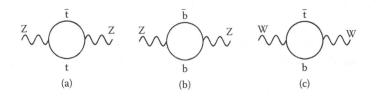

FIGURE 22.1 Feynman diagrams representing the propagation of an electron. The first term corresponds to the free electron, and the second to the first perturbation. There are infinitely many higher-order terms.

FIGURE 22.2 Loop corrections to W and Z boson propagators, involving the bottom and top quarks.

The current value, obtained by measuring the top quark mass directly, is about 173 GeV. A remarkable agreement!

Because we know the mass of the top quark, we can now consider smaller loop contributions to $\Delta\rho$. It turns out that this probes the Higgs, because the Higgs contributes to the mass of the Z boson through the diagrams in Figure 22.3. However, while the effects of the top quark grow with the square of its mass, the effects of the Higgs mass are much milder.* As a result, bounds on the Higgs mass from this source alone are much looser than they would be for the top quark. There are a number of other precision experiments, to which the Higgs also makes a mild contribution.

Most of these experiments were performed during the 1990s at SLAC's SLC and at CERN's Large Electron-Positron (LEP) facilities. Their competitive/collaborative efforts yielded results of unprecedented precision, known collectively as precision electroweak experiments. These facilities competed to get the best results, faster. But they also collaborated intimately, the ideas from one quickly influencing the other. In fact, Burt Richter, who was the director of SLAC during most of this time, was also the first to propose the construction of LEP, during a sabbatical leave at CERN in 1976. That science advances through such a fine equilibrium between competition and collaboration may be something quite difficult for nonscientists to believe.

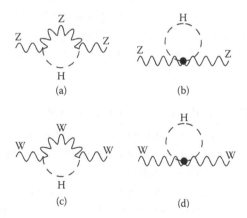

FIGURE 22.3 Loop corrections to the W and Z boson propagators, involving the Higgs.

---

* They grow only logarithmically with the Higgs mass.

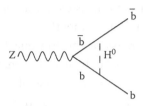

FIGURE 22.4   Diagram contributing to the decay of the Z into a b quark and a b antiquark, which involves the Higgs boson.

The experiments at the SLC and LEP collided electrons with positrons so that the total energy was equal to the value of mc² for the Z boson. Thus, the Z boson was produced at rest but immediately decayed since its lifetime is less than a billionth of a billionth of a second. The decays were into pairs of leptons (e⁺e⁻, μ⁺μ⁻, τ⁺τ⁻) and pairs of quarks. The detailed measurements of these decays could also be explained in terms of the same parameters; again, virtual effects had to be considered, including those involving the Higgs boson, as shown in Figure 22.4.

Combing the precision experiments at LEP and SLC with precision determinations of the masses of W, Z and the top quark leads to tight constraints on the fundamental parameters of the standard model and also an indirect constraint on the Higgs mass. The conclusion is that with 95% confidence the Higgs mass should be less than 166 GeV. As discussed below, the direct search for the Higgs boson at LEP excludes a mass less than 115 GeV. The important point to emphasize is that the data cannot be fit without including the virtual effect of the Higgs boson. If the standard model Higgs boson is not found with an appropriate mass, then something else must be added to the standard model.

22.2 DETECTING THE HIGGS BOSON DIRECTLY

The Higgs mechanism is the source of the mass for all particles, whether quarks, leptons, or weak gauge bosons. The associated Higgs particle couples to these particles proportionally to their masses. A lower mass implies a smaller coupling; a higher mass implies a larger coupling. Particle accelerators start out from electrons, protons, and their antiparticles. Since protons are made of the light quarks u and d, the particles accelerated at colliders are light, and thus couple weakly to the Higgs boson. As a result, the production of the Higgs boson involves always the virtual effects of heavy particles.

There are many higher-order effects involved in Higgs production. Which one is most important depends on the exact characteristics of the machine under consideration (which particles are accelerated and which is the energy of the collisions) and also on the actual value of the Higgs mass.

Once the Higgs is produced, one must consider how it decays. Unfortunately, many decays can mimic the signals from a Higgs decay. This is known as background, and it might be much larger than the signal from the Higgs decay. A huge effort involving theorists, predicting what might happen, experimentalists, detailing what might be measured given the detector capabilities, and a very intensive computer simulation program is needed in order to increase the signal-to-background ratio. This varies from one facility to another and depends on the Higgs mass. Sometimes information associated with a particular production mechanism might also influence how well a specific decay is measured. We will now concentrate on the current bounds on the Higgs mass from LEP and the Tevatron. Later, we will cover what might be obtained at Large Hadron Collider (LHC) and the proposed International Linear Collider (ILC).

Once a Higgs boson is produced it will decay very quickly. Since the Higgs bosons are coupled to particles proportional to their masses, the dominant decay modes one can think of would be $H \to b\bar{b}$, $H \to \tau^+\tau$, $H \to W^+W^-$, and $H \to ZZ$.* The top quark is so heavy that direct decay to $t\bar{t}$ is not possible. If the Higgs mass is below 165 GeV, the direct decays to WW and ZZ are not energetically possible (remember that mc^2 for the W is 83 GeV), but a decay is possible into a real W and a virtual W. The closer the Higgs mass gets to 165 GeV, the more important this mode is, because it becomes easier to produce the virtual W.

Branching ratios (probabilities) for different decay modes are shown in Figure 22.5 for various masses of the Higgs boson, m_H. For a mass of 115 GeV near the present limit the dominant decay mode is to $b\bar{b}$. However, this mode is not easy to detect at a proton-proton collider because of the background. In fact, for such a low mass an important decay mode may be the decay into two gamma rays. Even though this decay is only 2 out of 1000 decays, the signal of two high-energy photons with a total energy of order 115 GeV would be very unique. For larger masses the WW mode becomes increasingly important.

---

* Recall that a bar over a particle represents its antiparticle. Thus, $\bar{b}$ represents the bottom antiquark and $\bar{t}$ represents the top antiquark.

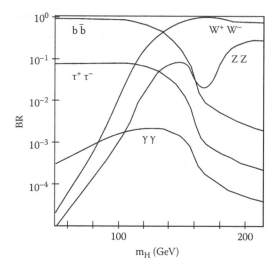

FIGURE 22.5 Branching ratios (BR) for several Higgs decays as a function of the Higgs mass, m_H.

22.3 DIRECT HIGGS SEARCHES AT LEP

LEP was a circular particle accelerator built along a 27 km long tunnel situated at CERN. CERN was born in the early 1950s, in the wake of the war, with the objective of establishing a world-class scientific laboratory probing the inner secrets of matter's smaller constituents. Emphasizing its international nature, it encloses sections of the French-Swiss border, lying partly in one country, partly in the other. LEP was constructed between 1985 and 1989. Excavating its 27 km tunnel was a major civil engineering feat; getting the accelerator to work was a major electrical engineering achievement; setting up its four gigantic detectors (named ALEPH, DELPHI, L3, and OPAL) required the best minds from science and technology. Such enormous challenges were only surmounted due to the truly international effort that went into LEP. LEP finally shut down in 2000 to allow for the construction of LHC.

Although the facility was colossal, only very small errors were tolerated. The energy of the beam and its position were known so precisely that they could be used to trace tidal effects that alter the circumference of the ring by 1 mm the level of the water in nearby Lake Geneva, and even the passage of the TGV fast trains on the tracks nearby. Can you imagine sitting in a meeting to discuss alterations detected in the beam of elementary particles, standing up, and stating that the effect was due to the moon?

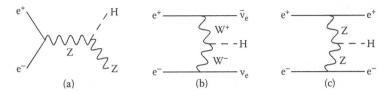

FIGURE 22.6 Possible mechanisms for production of a Higgs boson at LEP. (a) Associated production, also known as Higgs-strahlung, is the dominant contribution. (b) WW fusion and (c) ZZ fusion give minor contributions.

At LEP, electrons collided with positrons so that the total energy was 93 GeV, equal to mc^2 for the Z boson. LEP produced over 17 million Z bosons. After an upgrade, the energy was increased, allowing further probes on the nature of W and Z bosons and their interactions, and to search for the production of the Higgs boson.

The production of the Higgs boson at LEP should occur mainly through the process in Figure 22.6a, with the processes in Figure 22.6b and c providing a smaller contribution. One searches for the Higgs mainly through its decays into a b quark and a b antiquark. For a Higgs particle of 115 GeV this decay occurs around 74% of the time. In turn, the Z can decay into quarks, leptons, or neutrinos. A configuration is also searched for in which the Higgs decays into tau and antitau, while the Z decays into a b quark and a b antiquark, but this is much rarer since a 115 GeV Higgs only decays about 7% of the time into this final state. When one combines the results for all possible final states obtained at all four LEP experiments, a lower bound can be placed on the Higgs mass: with 95% confidence the mass must be larger than 115 GeV.

22.4 DIRECT HIGGS SEARCHES AT THE TEVATRON

The Tevatron is a facility situated at Fermilab, outside Chicago, colliding protons with antiprotons. After LEP's results were known, the Tevatron was upgraded, starting what is known as RunII, at an energy of 1.96 TeV.

Because protons and antiprotons have quarks, the Higgs production mechanisms are different from those possible at electron-positron machines such as LEP. With six quarks in the initial state, the strong interactions will easily produce gluons. Calculations indicate that, as a result of the large Higgs coupling to the top quark, it is the gluon-gluon fusion shown in Figure 22.7 that is the dominant production mechanism. The next relevant production mechanisms, in decreasing order of importance, are the associated production with a W and the associated production with a Z, shown collectively in Figure 22.8a. Next comes the WW fusion and ZZ fusion processes, shown collectively in Figure 22.8b. For Higgs masses below about 140 GeV, the production mechanisms of Figure 22.8a dominate those in Figure 22.8b. As the Higgs mass increases, the importance of the mechanism in Figure 22.8b increases. Figure 22.6 is analogous to Figure 22.8, except that the former refers to the electron-positron in the initial state, while the latter refers to the quarks within the proton and antiproton in the initial state.

For Higgs masses close to those probed at LEP, the largest decay rates of the Higgs particle continue to arise from its decay into a b quark and a b antiquark ($b\bar{b}$) or from its decay into a tau-antitau pair ($\tau^+\tau^-$). However, for Higgs masses above 140 GeV, the Higgs can decay into a W^+W^- pair*, which becomes the dominant decay channel, and into a ZZ pair.

Unfortunately, in a hadronic environment there are many background events mimicking the $b\bar{b}$ pair coming from a possible Higgs decay. The situation for the gluon-gluon fusion production mechanism becomes difficult for low Higgs masses. Thus, it is difficult for the Tevatron to search for a Higgs boson with a mass less than 140 GeV. Fortunately, if the Higgs is

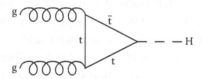

FIGURE 22.7 Gluon-gluon fusion production mechanism for the Higgs boson. This is the dominant production mechanism at hadron colliders.

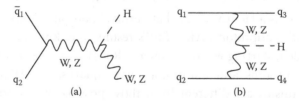

FIGURE 22.8 Subdominant mechanisms for Higgs boson production at hadron colliders: (a) associated production with a gauge boson (which may be a W or a Z) and (b) WW fusion or ZZ fusion.

---

* If mc² for the Higgs boson is smaller than $2m_Wc^2$ (for the W^+W^- pair), then one W is a virtual particle.

produced in association with a W, as in Figure 22.8a, we can use its presence to get rid of many background events. Said otherwise, a $b\bar{b}$ event coming from a Higgs decay is hard to disentangle from the background. But a $b\bar{b}$ event in association with a W, as would occur for a Higgs produced through the diagram in Figure 22.8a, is easier to set apart from a mere background event. The Tevatron also searches for Higgs decays into a tau-antitau pair. This decay can be used in connection with all production mechanisms.

If the Higgs boson has a mass greater than 140 GeV and is produced from gluon-gluon fusion (Figure 22.7), then the decay into W^+W^- can be detected. This is the best opportunity for finding the Higgs boson at the Tevatron. Combining all searches from the two experiments CDF and D0, the Tevatron might yield significant information about the Higgs if its mass is between 140 and 190 GeV by the end of 2010, when it is expected that the experiments will end. In early 2009, the Tevatron found with 95% certainty that the Higgs cannot have a mass between 160 and 170 GeV.

NOTE

1. Numbers quoted by the DZero top quark discovery article published in *Phys. Rev. Lett.*, 74, 2632–37 (1995).

CHAPTER 23

What We Hope to Learn from the LHC

23.1 WHAT IS THE LHC?

LHC is the acronym for Large Hadron Collider. It is a circular facility, built at CERN in the 27 km long tunnel where the Large Electron-Positron (LEP) used to be. It is used to accelerate protons to 99.9999991% of the speed of light. Two beams of protons move in separate high-vacuum pipes before they are made to collide (hence the name) with each other. The collisions occur at four locations, one for each of the main detectors: ATLAS, CMS, LHCb, and ALICE. The first two will be used to look for the Higgs. There will be around 600 million collisions per second, each collision with an energy of 14 TeV. This is about seven times the energy attained at the Tevatron, and the intensity of the beams (known as luminosity) will be about 40 times larger.

In order to curve, the protons are subject to magnetic fields produced by 9,300 very large magnets. Because the protons move so fast, the magnetic fields must be extremely intense. Their production requires extremely large electric currents and low electrical resistances. Special superconducting magnets working at the extremely low temperature of −271°C (around −454°F) had to be built. This requires the world's largest liquid helium cooling system. There is no room for mistakes. If a problem occurs, one will have to heat things up to room temperature, make the appropriate repairs, and then cool things back down. This is exactly what happened shortly after LHC was turned on, provoking a delay of more than a year.

The CERN release described "a faulty electrical connection between two of the accelerator's magnets. This resulted in mechanical damage and release of helium from the magnet cold mass into the tunnel."[1] The skill required of every worker involved is truly amazing.

Because there are about 600 million collisions per second, the amount of data taken exceeds anything attempted before. Although most events that do not look promising are discarded at the detector level (through so-called triggers) and not recorded, CERN quotes roughly 15 million gigabytes of data to be produced annually. This is enough to fill a hundred thousand DVDs a year! It is estimated to be about 1% of the world's information production rate. Many scientists wish to access this huge amount of data. Therefore, a worldwide network of computer sites has been set up, known as the Grid. The idea is to share computer power and data storage over the Internet.

ATLAS is about 45 m long, 25 m wide, and 25 m high. It weighs about 7,000 tons. CMS is about 21 m long, 15 m wide, and 15 m high. It weighs about 12,500 tons. You may think of these as ten-story-high, long buildings. But they have almost no empty room. They are jam packed with the latest in technology in order to detect elementary particles and track their motion. And some components must be placed with submillimeter precision and withstand radiation levels much larger than those felt by the communication satellites in orbit. That LHC and its experiments could be built is a testament to human ingenuity.

23.2 HIGGS SEARCHES AT LHC

Regardless of its precise mass, a standard model Higgs boson will be produced at LHC predominantly through the gluon-gluon fusion mechanism of Figure 22.7. In contrast, the Higgs decay that will allow its detection depends very strongly on the Higgs mass. For Higgs masses up to 140 GeV one would like to use the $H \to b\bar{b}$ decay. Unfortunately, this is extremely difficult to disentangle from background events coming from a trivial production of b quarks in the collision of the quarks within the protons, driven by the strong interaction, and thus completely unrelated to the Higgs particle. As a result, one must look at the Higgs decay into two photons. Since this is a rare event, low Higgs masses might be the hardest to detect.

For Higgs masses between 140 and 180 GeV, the best detection mechanism comes from the Higgs decay into two W bosons, one real and one virtual. Each W is detected through its decay into a charge lepton and a neutrino. For a Higgs mass larger than 180 GeV, the Higgs decay into two

Z bosons becomes experimentally relevant, with the Z bosons detected through their decays into leptons.

The WW fusion and ZZ fusion diagrams in Figure 22.8b are known collectively as vector-boson fusion diagrams. While this production mechanism is about a factor of 10 smaller than gluon-gluon fusion, it is the second most important mechanism at the LHC. Furthermore, the Higgs is accompanied by two quark jets (see Figure 22.8b), which can be detected and help to reduce the background. This should make it possible to detect the decays $H \to \tau^+\tau$ and $H \to W^+W^-$ (with the W bosons detected through their decays into leptons), and possibly other decays for Higgs masses around 140 GeV or greater.

The separation of production mechanisms may help in keeping our ideas straight. But things are usually more complicated. Imagine that one looks for a Higgs decaying into a tau-antitau pair in association with two jets in the final state. This can be achieved by all production processes discussed above: gluon-gluon fusion, vector-boson fusion, and associated production with a vector boson. These are shown in Figure 23.1a to c, respectively. Several energy and momentum properties of the two jets and of the tau-antitau pair depend on the diagram in question. These properties may be used to disentangle one diagram from the next. For example,

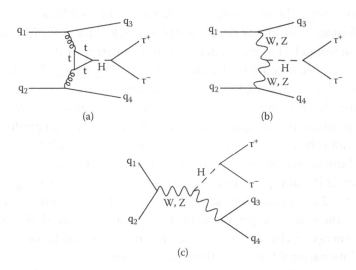

FIGURE 23.1 Feynman diagrams representing the production of a Higgs in association with two jets. The Higgs may be detected through its decay into $\tau^+\tau$. The diagrams represent (a) one gluon-gluon fusion diagram, (b) the vector-boson fusion, and (c) the associated production with a vector boson.

one might try to isolate the contributions from Figure 23.1b. This would probe the couplings of the Higgs particle to the weak gauge bosons.

Identifying the best methods to disentangle contributions is extremely difficult; it involves very extensive theoretical calculation, very intensive computer simulations, and a perfect control over what is admittedly a very complicated detector. One expects levels of 10% to 30% accuracy on the couplings HWW, Htt, and Hττ.

23.3 DISASTER? IF THE LHC CANNOT FIND THE HIGGS BOSON

Given the success of the standard model, there is great confidence that the LHC will discover the Higgs boson, if the Tevatron has not done it already. If, however, the Higgs boson is not found, it would seem a terrible failure after all the effort and expense. The truth is that most physicists would be more excited by the failure than by the success. If the Higgs boson remains undetectable at the LHC, this means that there must be physics beyond the standard model: that would be the really important discovery.

There are two general possibilities for new physics that would affect the search for the Higgs boson; either or both could lead to the failure at the LHC. The first possibility is that the decay might be primarily to other new particles not yet discovered. One possibility would be the decay into a pair of fairly light neutral particles that are either stable or have a very long lifetime. Thus, all the decay channels searched for at the LHC or the Tevatron would be too rare to detect. This is sometimes called the hidden Higgs. Such decay could be detected by the missing energy, but this may be impossible in the complicated final states at the LHC.

At the electron-positron collider LEP, a careful search was made for events in which the positron combined with the electron to produce a Z boson and missing energy. This was possible because at LEP you do not have all the debris that exists at the LHC from the strong interactions. In this way LEP could put a limit on the mass of the hidden Higgs boson of about 112 GeV, practically as good as the limit for the case of a visible decay. If the mass is larger, the hidden Higgs may remain hidden until a higher-energy electron-positron machine comes into existence, the proposed International Linear Collider (ILC) accelerator.

The other possibility is that the role of the Higgs is not played by a single particle. Some possibilities are discussed in the next chapter. If there are several Higgs particles, it is possible that a signal of one of them could be discovered at the LHC, but it might be difficult to interpret the signal because it did not have the characteristics of the Higgs boson of the standard model.

23.4 IF THE LHC DISCOVERS THE HIGGS BOSON, IS THAT THE END OF THE STORY?

The announcement of the discovery of a Higgs boson at the LHC will lead to great excitement in the physics world and headlines such as "Physicists Discover the Missing Link." Having found this last missing particle, can the physicists rest in peace knowing that the beautiful standard model has been confirmed? No! Once a Higgs boson is discovered, the work begins to study this new particle to see if indeed it has just the properties required by the standard model.

If the Higgs boson has a mass near 115 GeV, we expect it to be discovered by its decay into two photons, having been produced by gluon-gluon fusion. This decay cannot occur directly. In the standard model it is due to loops involving W bosons and quarks. The most important standard model contributions are shown in Figure 23.2. As can be seen by Figures 23.2 and 22.7, both the production and decay depend on the coupling of the Higgs to the top quark and top antiquark (usually represented by Htt). The rate of this observation can be used to determine the coupling Htt of the Higgs to the top quark as well as the coupling to the W, but there will be considerable uncertainty since it depends on calculating how the gluons are produced by the colliding quarks. One hopes to be able to determine the Htt coupling in this way with an accuracy of 10% to 20%. However, this gives us no information about the Higgs coupling to the other quarks or to the leptons.

Once the Higgs boson is discovered and its mass is determined, it may be possible even for this low mass to detect the decay into tau-antitau and obtain a rough idea of the Hττ coupling. However, it will be difficult to

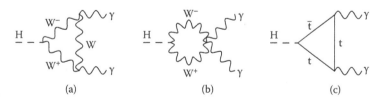

FIGURE 23.2 Feynman diagrams representing the decay of a Higgs into two photons. The dominant contribution comes from diagrams (a) and (b) involving W bosons; diagram (c) with the top quark gives a smaller contribution with the opposite sign.

determine the couplings to the vector bosons, and there will be no evidence of the coupling to the b quarks.

If the Higgs boson has a large mass of around 160 GeV, we expect to discover it by the decay into WW, with the production vector-boson fusion (Figure 22.8b). This will give information on the Higgs coupling to the vector bosons but not about the coupling to quarks and leptons. Again, knowing the mass, it may be possible to identify the decay into tau-antitau. Determining the Hττ coupling would require disentangling different diagrams, like those of Figure 23.1, which requires extensive theoretical calculations as well as computer simulations of the collisions and the detector.

In any case, it is clear that the LHC will only give limited information on the interactions of the particle that it discovers. In particular, it will be very difficult to determine the Hbb coupling because, as noted, it is very difficult to uncover B mesons in the complicated final states produced in proton-proton collisions. In some theories with two Higgs bosons (instead of the single one present in the standard model), one Higgs boson couples to the b quark and a different Higgs boson couples to the t quark.* How can we tell such a theory apart from the standard model? One would need detailed information about the couplings to the b and t quarks. And it should prove much easier to obtain good information about the Higgs interactions from a high-energy electron-positron collider like the proposed ILC.

NOTE

1. CERN Press Release PR14.08, 16.10.2008, available online at http://public. web.cern.ch/press/PressReleases/Releases2008/PR14.08E.html.

---

* Incidentally, this could possibly explain why the mass of the bottom quark is so much smaller than the mass of the top quark.

CHAPTER 24

Possibilities for the Future

24.1 MULTI-HIGGS MODELS

We have mentioned that there is one gauge boson for electromagnetism, three for the weak interaction, and eight for the strong interaction. These numbers are set by the gauge symmetries, which determine the respective interactions. On the other hand, there is no known theoretical reason for the number of fermion families. Each family has two leptons (a charged lepton and a neutrino) and two quarks (a charge +2/3 quark and a charge –1/3 quark). And we know that that there are three families of quarks and leptons. Although we know of no theoretical reason that explains this fact, three families turn out to be the minimum required to accommodate CP violation into the Cabibbo–Kobayashi–Maskawa quark mixing matrix.

As happens with the fermions, there is also no theoretical constraint on the number of Higgs fields. So, why should there be only one Higgs particle? This is clearly the simplest solution, and thus it has been built into the standard model. But it is certainly not compulsory. Indeed, many believe that having only one Higgs would be highly unnatural. As a result, theories with several Higgs fields have been studied almost since the beginning of the standard model.

The most studied case adds only one extra scalar doublet to the standard model fields and goes under the long name of two-Higgs-doublet model. Already this simple model has many interesting properties that distinguish it from the standard model. The first is the presence of three neutral and two charged scalar particles, instead of the single Higgs particle that appears in the standard model. To understand why this occurs, we must look back at the standard model. Without the Higgs mechanism, the W⁺, W⁻, and Z bosons of the standard model would be massless, each with only two possible transverse polarizations. With the Higgs mechanism, W⁺, W⁻, and Z become massive, each with three polarizations (the two transverse and a new longitudinal one). How does this come about? The Higgs field has four components, two are charged and two are neutral. After spontaneous symmetry breaking, the gauge symmetry is hidden: the two charged Higgs components and one neutral component are absorbed as the longitudinal components of W⁺, W⁻, and Z bosons; the fourth (neutral) component corresponds to the Higgs particle. In two-Higgs-doublet models everything is doubled. There are four charged components. Two are absorbed by the charged W bosons, and two remain as a physical particle-antiparticle pair of charged particles with no spin.* This is the first difference with respect to the standard model. We get two charged scalar particles, while there were none in the standard model. But there are also four neutral scalar components. One is absorbed by the Z boson. Thus, there are three neutral scalar particles left, while there was only one in the standard model (the almighty Higgs particle). This raises the very exciting possibility that LHC will detect not one but five Higgs scalars, or possibly even more.

The presence of new particles would mean that they would contribute as virtual particles to known processes. If they exist, their properties are constrained. For example, barring any theoretical impositions, the presence of a second scalar doublet would allow a neutral Higgs to couple a d quark to an s antiquark. Such possibilities are known as flavor-changing neutral scalar currents. Flavor-changing neutral scalar currents would lead to the diagram in Figure 24.1, constituting a new contribution to KKbar mixing. Similar diagrams would mediate mixing in the B, B_s, and

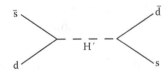

FIGURE 24.1   Flavor-changing neutral scalar currents mediating KKbar mixing at tree level.

---

* Recall that a scalar particle is a particle with zero spin.

D particle-antiparticle neutral systems. Because the mixings are small and are well explained by the standard model through loop effects, such new tree-level contributions must be very constrained. Either the masses of the new particles are very high or the couplings are very small (perhaps zero), or both. A possibility proposed by Paschos and independently by Glashow and Weinberg is to invoke a new discrete symmetry to couple one Higgs field only to up quarks, while the other Higgs field couples only to down quarks. Although they proposed exactly the same, and their articles appeared back to back in the same issue of the *Physical Review*, Paschos's article has around 270 citations while Glashow and Weinberg's (both Nobel Prize recipients) has around 940 citations. Science is, after all, part of the human experience.

Interestingly, two-Higgs-doublet models raise the possibility of spontaneous CP violation, as first pointed out by T. D. Lee. The idea is that the equations describing the model would be invariant under the CP transformation, but the vacuum of the theory would not. The vacuum expectation values of the two scalars would have a relative phase, thus permitting CP violation. In contrast, in the present standard model CP violation enters explicitly in the coupling of the Higgs boson to the quarks, producing the phase in the Cabibbo–Kobayashi–Maskawa matrix. This is known as explicit CP violation. It is a curious historical fact that these two possibilities—the explicit CP violation proposed by Kobayashi and Maskawa, and the spontaneous CP violation proposed by T. D. Lee—were both put forth in 1973.

Within two-Higgs-doublet models, it is difficult to reconcile spontaneous CP violation with the need to suppress flavor-changing neutral scalar currents. In 1976 Weinberg used a three-Higgs-doublet model to solve this problem. This model has four charged scalars, which come in two particle-antiparticle pairs. These can mix, and their mixture can be a novel source of CP violation. In addition, the model has five neutral scalars.

In general, there might be any number, N, of Higgs doublets. We just don't know how many there really are. This is an issue that LHC and the proposed ILC might begin to explore.

24.2 THE NEED FOR FURTHER TESTS AND THE IMPORTANCE OF ILC

Once the Higgs is detected, one must check its other properties: its decay width, the decay ratios into specific final states, the couplings to the gauge bosons, the couplings to the top quark, its self-couplings, its spin, parity, and CP quantum numbers, etc. Which properties one can measure depends crucially on the production mechanism, decay process, and Higgs mass.

As noted in Chapter 23, there are likely to be several Higgs couplings that cannot be determined at the LHC. Part of the problem is that the protons that collide at LHC are made of quarks and also have a far from trivial gluon content. A collision between two protons is really a very complicated mess where quarks and gluons from one interact with quarks and gluons from the other. The energies of the colliding partons (quarks and gluons) cannot be measured with certainty, limiting the information one can gather from the resulting interactions.

In contrast, electrons and positrons are fundamental particles and one can determine the energy of each collision very precisely. This provides a much better handle on studies of the particles produced in the collision, such as the Higgs particle.

ILC is the acronym for the International Linear Collider. It is a proposed linear accelerator, roughly 30 km long, which would involve the collision of an electron with a positron with a total energy of 500 GeV. The essential components would be two head-on linear accelerators, one accelerating electrons to 250 GeV and the other accelerating positrons to 250 GeV. Because it is likely to cost more than the LHC, this must be an intercontinent facility, with Europe, America, and Japan sharing the construction and the glory. If approved, the facility could start its operations by the 2020s.

At ILC, there is no gluon–gluon fusion. Higgs production occurs mainly through the same processes present at LEP, shown in Figure 22.6: the production of the Higgs in association with a Z, or its production through WW and ZZ fusion. This machine can truly test the couplings of the Higgs boson. The HWW and HZZ couplings will be measured to about 2% precision. The Htt, Hbb, Hcc, and Hττ couplings can be measured to an accuracy of a few percent. The standard model also contains an interaction between three Higgs bosons, a so-called HHH coupling. This could be determined with an accuracy of about 10% to 20%, thus providing another check on the theory of the Higgs boson.

Most importantly, such a machine would provide a window into the unknown. All through history, a new device has provided a new window into uncharted territories. Much like Columbus finding America on his search for India, we have always been thrilled by the events we could predict, but we were always amazed at the new things we encountered.

Thomson's cathode rays gave us the electron. Spectral studies hinted at the atom. Cosmic ray detectors gave us the positron, the muon, the pion, and the strange particles, which specially built particle accelerators turned into a zoo. Each new energy barrier brought new puzzles and discoveries: the various quarks, the third family of fermions, W and Z gauge bosons. In the last 10 years alone we have found CP violation in new systems, we realized that neutrinos do have a mass, and we are on the verge of probing the origin of mass itself. Who knows what wonders LHC and ILC will uncover?

CHAPTER 25

Conclusion

The standard model of elementary particles was fully formulated by 1973. Experiments in the next 10 years provided important evidence in favor of it. These included the discovery of neutral current interactions in 1973, the discovery of the b quark in 1977, and the discovery of the W and Z bosons in 1982 and 1983. In the following 25 years many more experiments all over the world provided precision tests of the theory.

There are many reasons, however, to believe that there is more to be discovered. There is one direct observation that is inconsistent with the standard model: the evidence for nonzero neutrino masses. New experiments to study neutrino masses and mixings are being planned in many different countries. In order to probe whether neutrino mass is associated with a violation of lepton number, experiments using large masses of several different isotopes are being planned to search for neutrinoless double beta decay.

The standard model of CP violation given by the Cabibbo–Kobayashi–Maskawa (CKM) matrix is consistent with experiments done so far, but these do not provide precise tests of the theory. If there is new physics at a somewhat higher mass scale, we would expect it to have CP violation and, via virtual effects, produce small but significant deviations from the standard model predictions. Further studies of CP violation involving mesons containing b quarks will take place at the electron-positron collider in Japan and with the LHCb detector at CERN. The possibility of CP violation in neutrino mixing will be explored with the new neutrino experiments.

There remains one particle in the standard model that has not been detected, the mysterious Higgs boson. While the success of the theory so far requires that something play the role of the Higgs, there is no reason why it is just a single particle. After all, we have six quarks—why only one scalar boson? A major goal of the large new accelerator, the LHC, at CERN in Geneva is to discover the Higgs boson or whatever it is that plays the part. Even if a candidate for the Higgs boson is discovered at the LHC, it may be difficult to study its interactions and demonstrate that it serves all the functions required of the Higgs, which is to give masses to all the massive particles. That is one argument for a new high-energy electron-positron collider, the International Linear Collider (ILC), where the Higgs boson could be studied without the large background resulting from the strong interactions at the LHC.

A wonderful goal of science starting with the work of Newton is to understand the universe around us using the same laws of physics that hold here on earth. A major problem that has arisen is the evidence that most of the matter that gravitationally holds galaxies together does not consist of the matter that we have discovered here on earth. The general belief is that this dark matter consists of some particles that were produced in the early days of the universe, referred to as weakly interacting massive particles (WIMPs). The hope is that these particles may be produced by the high-energy collisions at the LHC. Another cosmological question is the reason that our universe is made of matter and not antimatter; the suggestion by Sakharov was that this was due to CP violation in the early days of the universe. However, the CP violation of the standard model is not sufficient for this purpose, which is one reason for interest in other sources of CP violation.

There are many theories that go beyond the standard model. A popular example is called supersymmetry, in which for every particle in the standard model there is a partner with a different spin. So far none of these extra particles have been discovered, but the lightest of them is a candidate for the WIMP. There are many articles and books about string theory and extra dimensions, but the experimental consequences of these are far from clear.

As we look back at the discoveries in the twentieth century, we see they involve a back and forth between experiment and theory. Some theories, like Pauli's neutrino, are attempts to explain a particular experimental problem; others, like Weinberg's gauge theory of weak interactions, are primarily motivated by theoretical arguments. Some experiments are designed to test theories, like Rubbia's search for W and Z bosons, while others involve careful exploration of some phenomena, like the discovery of strange particles in the cosmic rays. As we look ahead in the twenty-first century, we cannot tell you what the next great discovery will be, but we can be sure it will involve an exciting intersection of theory and experiment, as in the history we have presented here.

Appendix 1: Important Twenty-First-Century Experiments

This is a list of some of the important experiments that are being prepared in the first decade of the present century, many of which should produce exciting results in the second decade. The goals of these experiments are discussed in Parts B, C, and D.

A1.1 LARGE HADRON COLLIDER (LHC)

Protons will be collided against protons at an energy around one order of magnitude greater than in any previous experiment. Two huge detectors, CMS and Atlas, will analyze the set of particles produced by the collision. The hope is to discover evidence of fundamental particles never detected before. In particular it is expected to discover the Higgs boson, the missing link in the standard model. First results may come in the next few years.

A1.2 INTERNATIONAL LINEAR COLLIDER (ILC)

This is a proposal to collide electrons with positrons at an energy five times larger than the previous collider, large electron-positron (LEP). A major goal would be to study any new particles discovered at the LHC without the problem of the large background that exists at the LHC due to the strong interactions. The decision on whether and when to build the ILC may depend on the results from the LHC.

A1.3 KEK SUPER B FACTORY

This is an electron-positron collider designed to produce B mesons. It represents an upgrade by a factor of 10 to 100 in intensity from the previous experiment at KEK in Japan. The goal is to make more and more precise measurements of CP violation to detect possibilities of physics beyond the standard model. It is expected to start in 2012.

A1.4 LHCB

This is a specially designed detector at the LHC that will study B and B_s mesons produced by proton-proton collisions. This will have the same goal as the KEK B factory but will be able to study CP violation in B_s meson mixing. Also, given the very large number of B mesons, it will be possible to search for rare decays.

A1.5 JAPAN PROTON ACCELERATOR RESEARCH COMPLEX (JPARC)

A new high-intensity proton accelerator in Japan will be used to produce a large number of K mesons. One goal will be to look for extremely rare decays that can be used to test the standard model of CP violation.

A1.6 SEARCH FOR THE ELECTRIC DIPOLE MOMENT OF THE NEUTRON

The neutrons from the Spallation Neutron Source at Oak Ridge in the United States will be confined in a large tank of liquid helium exposed to large magnetic and electric fields. The goal is to test for an electric dipole moment of the neutron with a sensitivity one hundred times greater than in any previous experiment. A nonzero value would be a signal of time-reversal violation inconsistent with the standard model of CP violation.

A1.7 SEARCH FOR NEUTRINOLESS DOUBLE BETA DECAY

A number of experiments with different isotopes are planned to look for decays that emit two electrons and no neutrinos. These would be located in underground laboratories in Italy, Canada, and Japan and in a large new underground laboratory under construction in the United States. The goal is to discover the new physics responsible for neutrino mass.

A1.8 REACTOR NEUTRINO OSCILLATIONS

Experiments designed to detect the disappearance of antineutrinos at a distance of 2 km are in progress at the CHOOZ reactor in France and the Daya Bay reactors in China. The goal is to determine the magnitude of the one element of the neutrino mixing matrix that so far has proved too small to measure; this is the element required for CP violation.

A1.9 LONG BASELINE NEUTRINO OSCILLATIONS

Experiments are now in progress sending beams of muon neutrinos from the new accelerator in Japan to SuperKamoikande, from Fermilab to Minnesota, and from LHC to Grand Sasso. There are plans for experiments with still larger baselines, longer than 1,000 km. By studying the dependence of oscillations from ν_μ to ν_e on distance and energy, it is hoped to measure possible CP violation in neutrino mixing analogous to that in quark mixing.

A1.10 ICECUBE AND KM3NET

These are experiments in which a cubic kilometer of water is instrumented to detect high-energy neutrinos from outer space. IceCube is located at the South Pole and the water is in the form of ice, while KM3NeT is located in the Mediterranean Sea. One goal is to supplement observations of gamma rays from energetic astronomical sources; another is the possible detection of neutrinos produced by the annihilation of dark matter.

Appendix 2: Renormalization, Running Coupling Constants, and Grand Unified Theories

In Feynman diagrams, a free electron propagating through space is represented by a straight solid line, while a free photon is represented by a wiggly line. This would be the whole story if there were no interactions anywhere. But there are indeed interactions in the world, and what we mean by a particle that has been detected is a particle that lives in the real interacting world and not some abstract "free particle." As a result, the particle lines and the vertex describing their interaction suffer corrections due to the diagrams in Figure A2.1a and b.

Let us look at the first diagram. We introduced Feynman diagrams with the rule: "draw a straight line for each electron." But now it seems that this is not enough when interactions are taken into account. We also get the term in Figure A2.1a. So, an electron is instead represented by the sum in Figure A2.2. Thus, in the presence of interactions there is no such thing as free electrons. Because the quantum fields allow the creation and annihilation of photons, the interacting electron is always surrounded by a swarm of virtual photons, continually appearing and disappearing from the vacuum. We say that the physical interacting electron is dressed by the photons around. In contrast, we refer to the free electron as the bare

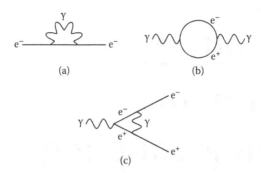

FIGURE A2.1   Feynman diagrams representing additional effects: (a) electron self-energy, (b) photon self-energy, and (c) vertex corrections.

FIGURE A2.2   Feynman diagrams representing the propagation of an electron. The first term corresponds to the free electron, the second to the first perturbation, and the next to some higher-order term. There are infinitely many such terms.

electron. How shameful! One may think of the electron as a juggler that is continually sending photons up in the air and recapturing them.

In the connection between Feynman diagrams and mathematical rules, the first term in Figure A2.2 is connected to the propagation, and hence to the mass, of the bare electron. We denote the mass of the bare electron by m_0, where the 0 subscript refers to bare or noninteracting. Since the propagator gets affected by the interaction, the electron's mass also gets affected by the interactions. A physical electron has a physical mass, m, not the bare mass, m_0, with which we started out. Figure A2.2 tells us how to relate the physical mass to the bare mass. The problem is that we have no access to the bare mass; it cannot be measured. Indeed, there is no such thing as a noninteracting electron. We can only measure the physical mass. But the calculations corresponding to the diagram in Figure A2.3 are performed with the bare mass. So, in real calculations one must perform the following three steps: (1) use diagrams like those in Figures A2.3 and more complicated ones in order to calculate physical processes in terms of the (immeasurable) bare mass m_0; (2) use Figure A2.2 in order to relate the bare mass, m_0, to the physical mass, m; and (3) combine the

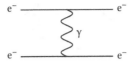

FIGURE A2.3 Feynman diagram representing the interaction between two electrons.

two calculations to express the physical processes in terms of the physical mass, m. This procedure is known as renormalization.

Naturally, Figure A2.1b states that a similar problem exists with the photon. The photon may create an electron-positron pair, reabsorbing it later on. Because the photon carries the electromagnetic interaction, which depends on the electric charge, this diagram implies a charge renormalization. That is, we write our theory in terms of a bare charge, e_0. This is what appears in the diagrams of Figures 4.1 and 4.2 of Chapter 4. But Figure A2.1b induces a change from the bare charge into the physical charge, e. Combining both calculations, we end up writing the physical processes in terms of the physical charge, e.

Including the relations between the physical and bare masses and charges, Figure A2.1c entails a calculable vertex correction whose effects can be measured experimentally. The precise agreement between theoretical calculations of fine effects like these and their experimental verification is perhaps one of the biggest success stories in all of science.

Renormalization is needed because we develop our initial calculation rules with noninteracting particles, while they live in an interacting world. It seems simple. But historically this was a major stumbling block. The reason is that the mathematical calculation of the diagrams in Figure A2.1 yields infinity! When calculating the probability for some physical process to occur, we perform a perturbative expansion with the following rationale: we calculate terms successively more complicated, but each new calculation involves a further power of the electron charge. If this charge is small, then each new term is smaller than the previous. For example, the vertex correction in Figure A2.1c should be much smaller than the vertex in Figure 4.2. But the alleged small correction is infinite! This bewildered many scientists.

It turns out that renormalization provides the solution. In quantum electrodynamics all infinities can be hidden in the relation between the bare parameters (e_0 and m_0) and the physical parameters (e and m). This is

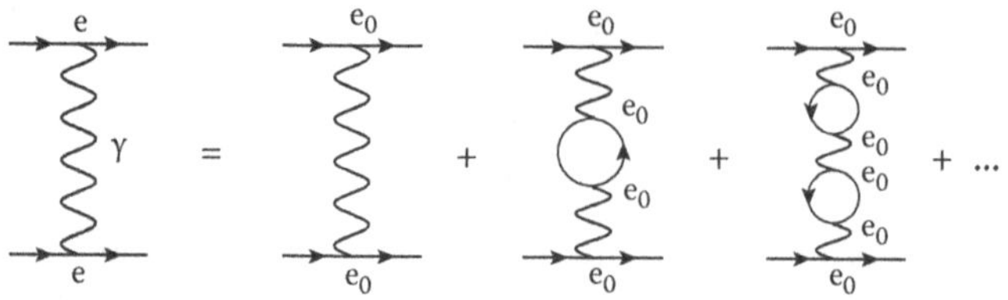

FIGURE A2.4   Relation between the physical charge, e, and the bare charge, e_0.

irrelevant for precise predictions, since we can't measure the bare parameters anyway; they have no physical meaning. But to many, this felt a bit like hiding an elephant under the rug. Even Feynman, whose scientific fame arose in great part from the development and use of this technique, suspected that the theory was not mathematically sound. Regardless of the philosophical interpretation of this procedure, one thing is sure: this yields the most precise predictions humankind has ever made. Whatever new interpretations or even mathematical advances are made, they will include something equivalent to renormalization.

In practice, there are several ways in which one can implement the renormalization procedure. One can say that the charge renormalization depends on the energy scale that the experiment under study is performed at. Take the tree-level diagram in Figure A2.3. We already know that the result depends on the square of the charge, e_0, and that we must also consider higher-order diagrams. The charge, e, measured by experiment arises from the effect of all these diagrams, as shown schematically in Figure A2.4. This is the charge renormalization we have already mentioned and which depends on the energy scale of the experiment. Thus, the charge measured at one energy scale will not coincide with the charge measured at another energy scale. One says that the charge "runs" with the energy scale.

There is a connection between energy and distance scale that can be used to gain an intuitive understanding of this effect. Consider the electron's wave. A wave can be used to probe an object if its wavelength is similar to the object's size. If the wavelength is too large, it will vary very little as it passes the object, and thus it will not probe it. If the wavelength is too small, it will probe the object's inner structure. If the wave is associated with a particle, de Broglie's relation tells us that the wavelength is inversely proportional to the particle's momentum. Thus, larger momentum probes smaller distances, and smaller momentum probes larger

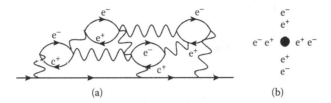

(a)   (b)

FIGURE A2.5   (a) A high-order loop diagram leading to charge renormalization, and (b) its intuitive interpretation as a polarization of the vacuum.

distances. To simplify, we will talk of a relation between energy and distance: experiments at small energy probe large distances; experiments at higher energy probe smaller distances.

Imagine that we use an electron to probe another electron, through a virtual photon, as shown in Figure A2.3. We know that a physical electron is not the same as a bare electron. In Figure A2.2, a bare electron is represented by the left-most straight line, while the physical electron arises from the combination of all higher-order diagrams. In complicated diagrams, there will be many loops with electron-positron pairs, as in Figure A2.5a. In each pair, the positron will tend to lie closest to the original electron. This can be viewed as the situation of Figure A2.5b. We say that the vacuum is polarized around the electron. In an experiment at low energies the virtual photon is shielded from the electron by a cloud of electron-positron pairs; thus, it sees a small charge. In an experiment at high energy, the photon will come deep within the cloud around the electron; the shield is not so effective, and it sees a larger charge. Thus, the electromagnetic coupling constant (roughly the square of the physical charge) increases as the energy increases. This is shown schematically in Figure A2.6.* Experiments at the electron-positron collider LEP measuring the annihilation rate into a muon-antimuon pair determined the electromagnetic coupling, α, to be 1/128 at the energy of about 100 GeV, in contrast to the well-known value of 1/137 for α at low energies.[†]

In contrast, the coupling constant of the strong interaction decreases with energy. This is a consequence of quantum chromodynamics (QCD)

---

* At low energies the weak interactions are weaker than the electromagnetic interactions, despite the fact that, as shown in Figure A2.6, the corresponding coupling constant is larger than the electromagnetic coupling constant. This occurs because the photon, which carries electromagnetic interactions, is massless, while the W and Z bosons, which carry the weak interactions, are massive. The lowest curve does not correspond exactly to the electromagnetic coupling constant, but it is closely related to it.

[†] In so-called CGS units, $\alpha = 2\pi e^2/(hc)$.

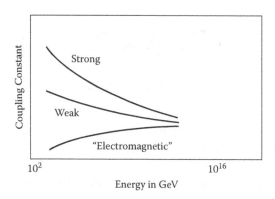

FIGURE A2.6 Schematic representation of the running of the coupling constants.

and is called asymptotic freedom. This occurs because the virtual gluons, the carriers of the strong force, have strong charges in contrast to the photon, which carries the electromagnetic interaction but has no electric charge. As discussed in Section 6.5, the smaller value of the strong coupling at high energy makes it possible to use perturbation theory to calculate high-energy processes like the production of jets. The increase of the strong coupling at low energies corresponding to large distances is the reason for the confinement of quarks.

An interesting observation is that the three couplings appear to approach the same value at a very high energy around 10^{16} GeV.* These curves assume that the only particles that exist and enter the loops are the ones we know already. Some theoretical physicists looking at these curves suggested that there was a single fundamental interaction at a high energy with a single coupling constant that was the basis for the three types of interactions we observe at low energies. This idea was labeled the grand unified theory (GUT).

As this theory was developed, it led to an interesting prediction that the proton should decay. An interesting decay predicted was into a positron and neutral pion. Large water Cerenkov detectors were built to look for proton decay. So far no proton decays have been detected; however, it was these detectors that discovered neutrino oscillations, as discussed in Part C.

---

* The three couplings are the strong SU(3), the weak SU(2) coupling g, and the U(1) coupling g'. At energies not far above the W and Z masses the electromagnetic coupling e = g' $\cos(\theta_W) = 0.87$ g'.

Appendix 3: Complex Numbers, Complex Fields, and Gauge Invariance

Complex numbers may be thought of in two different but related ways. We may think of them as arrows in the plane. Given a set of perpendicular axes in the plane, the arrow may be described by its components along the x axis and along the y axis (see Figure 14.1). We may also think of complex numbers as the combination of a garden-variety number (a real number such as 1, –1.1, or 3/5) and a dial. The dial says how much the arrow is rotated from the horizontal, and is known as the phase of the complex number. The real number indicates how large the arrow is, and is known as the magnitude of the complex number. The magnitude of a complex number V is usually represented by |V|. The relation between both representations is shown schematically in Figure A3.1. An arrow pointing to the right (whose phase is zero) corresponds to a positive real number; an arrow pointing to the left (whose phase is 180°) corresponds to a negative real number.

We add two complex numbers in the same way that we add two arrows. This is shown in Figure A3.2. The beginning of the second arrow is placed at the tip of the first arrow. The sum of the arrows goes from the beginning of the first arrow until the tip of the second. The order in which you perform the addition does not matter. You get the same result either way. The angle difference between the first arrow and the second arrow is known as the *phase difference*. Clearly, the size of the resulting arrow depends on this phase difference. If the two arrows had the same phase (were rotated

FIGURE A3.1   A complex number viewed as a number and a dial, or as an arrow. Here, for example, we show it as a dial multiplied by the number 2. The hand of a dial has unit length.

FIGURE A3.2   Sum of two complex numbers viewed as arrows.

FIGURE A3.3   Sum of two complex numbers with magnitudes equal to those in Figure A3.2, but with a different relative phase.

by the same angle) the size of the resulting arrow would be the sum of the sizes of the first and second arrows. If the phase difference were larger, as in Figure A3.3, then the size of the resulting arrow would be much smaller. A very special case occurs when we sum a complex number V with its negative, –V. The result is zero, as you would naively expect. This is such an important point that it is worth repeating. The size of the addition of two complex numbers depends on their relative phase. This is behind the phenomenon of interference, which we discussed in connection with waves. It is also central to quantum mechanics—the study of matter waves.

The product of two complex numbers is easier to understand in terms of the number-dial representation. An example is shown in Figure A3.4. The magnitude (the number part) of the product of two complex numbers is simply the product of the magnitudes of each. To find the direction of the phase (the dial) of the product of two complex numbers, we imagine that the dial is rotated by the phase of the first dial and then rotated by the phase of the second dial. Multiplying one complex number by a second number whose magnitude is 1 (which can be viewed as just a dial) changes the direction of the original arrow but leaves its magnitude unchanged. Multiplication by a dial (of magnitude 1) is the famous U(1)

FIGURE A3.4   Product of two complex numbers.

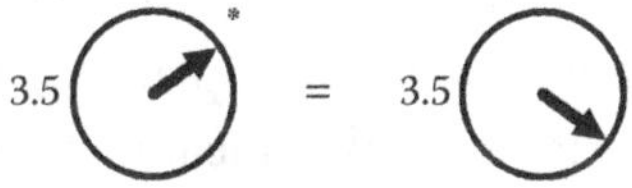

FIGURE A3.5   Complex conjugate of a complex number, indicated by *.

transformation, which Dirac asserts will leave the physical implications of the wave function unaltered.

The final operation needed to deal with complex numbers is the complex conjugation. This operation reflects the arrow about the horizontal. In the number-dial representation, it keeps the number (equal magnitude) and the dial appears rotated in the opposite direction (the phase gets a minus sign). This is shown in Figure A3.5. The complex conjugate of a complex number V is usually represented by V* (pronounced "V star").

We can combine the multiplication with complex conjugation. Multiplying a dial by its complex conjugate, we get a number that is equal to the square of the arrow's size; the magnitude squared. Symbolically, V V* = $|V|^2$.

The essential quantum concept discussed in Chapter 2 was that electrons, photons, etc., which we may think of as particles, propagate like waves. The question arises as to what is "waving": What is it that varies in time and space like a wave? Such a question arose in a different context when it was shown by Young in the early nineteenth century that light was a wave. The answer given then was that all space was filled with a mysterious substance called the ether, and it was the ether that was oscillating like water in a water wave. After the work of Maxwell and Einstein it became clear that there was no ether; what varied in time and space were the electric and magnetic fields.

Wave particles are specified by a quantum field: the symbol Ψ for the electron and A for the photon. The field Ψ is specified by a complex number at every point in space and time. The square of the magnitude is the probability of observing the particle at a given point. The variation of the phase in space and time is related to the wave property. This is illustrated in Figure A3.6.

FIGURE A3.6 Representation of a particle wave at five different points at the same time. The distance between the first and last picture is one wavelength.

In the experiment in which a wave particle passes through two equal slits and arrives at a point A on a screen placed some distance away, there may be destructive interference. This means that at point A the phase of the wave emerging from slit 1 is 180° away from the phase of the wave emerging from slit 2. For example, at one moment of time the dial for one may be at 90° (3 o'clock) and the dial for the other at 270° (9 o'clock). Since there is a single frequency governing the waves from the two slits, they remain out of phase at all times, and this is a permanent dark spot on the screen. It is worth emphasizing that the observation of interference shows us that electrons, photons, etc., propagate through the vacuum as waves; the use of complex numbers (dials or phases) has proven the most convenient way to represent this.

If we change the phase of the electron field by the same amount at every point in space, the physics is unchanged. If a particular wave function describes a certain electron, a wave function obtained from the first by rotating all dials by the same angle also describes the same electron. Figure A3.7 represents this central idea. The field of dials in Figure A3.7a describes the same electron as the field of dials in Figure A3.7b. This is referred to as global gauge invariance, or invariance under a global U(1) transformation.* There is also a field for the muon, which can decay by the weak interaction into an electron. In order for the physics to be unchanged, it is necessary to change the phase of the muon field by the same amount as the electron. More generally, all charged particles must be invariant under a phase change proportional to their charge. Associated with this global gauge invariance is the conservation of charge.

There is a more remarkable symmetry in quantum electrodynamics (QED) called local gauge invariance, or a local U(1) symmetry. In this case, the amount you change the phase by differs from one space point to the next. Figure A3.8 shows a field of dials obtained from those in Figure A3.7 by changing the angle (phase) of the dials at different locations by different amounts. Schrödinger's equation ascribes physical meaning to the relative angles between different dials. As a result, Figure A3.8 does

---

* The name gauge is a historical accident and has no relation to other uses of the word gauge.

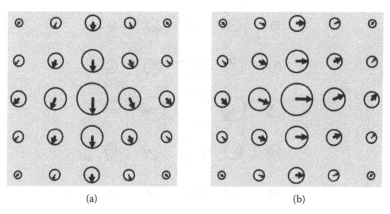

FIGURE A3.7 Two wave functions representing the same electron at a given time, as a function of space. The wave function in (b) is obtained by rotating all dials in (a) by the same angle of 90°. Such a global phase transformation does not affect the physical predictions.

not, in principle, describe the same electron as Figure A3.7. We are allowed to rotate all dials by the same amount, performing a global phase transformation. But we are not allowed to rotate the dials at different locations by different amounts, performing a local phase transformation. However, the possibility of local phase transformations becomes feasible when we take into account the way the electrons interact with the photons, in what is known as the *gauge principle*. This is a fundamental concept in particle physics.

To explain it, we must go back to the origins of the idea of a gauge transformation in classical electrodynamics. The electric field, E, is specified at any point by three numbers that give its components in each of the three space directions; the same is true for the magnetic field, B. Since each field has three spatial components, we need to know a total of six functions. However, as shown by Maxwell, E and B are not completely independent, so that it is possible to specify how E and B vary over space and time by the knowledge of just four functions of space and time: A^0, A^1, A^2, A^3. These are usually represented together as A^μ (where $\mu = 0, 1, 2, 3$), which is called the four-vector potential. In fact, there is an infinite (and continuum) number of possibilities for the choice of the function A^μ that yield the same values for E and B. The existence of this infinite number of possibilities became known as gauge invariance, a curiosity of classical electrodynamics.

Returning now to quantum theory, it remarkably turns out that, for any local change in the phase of the electron field, there is a corresponding

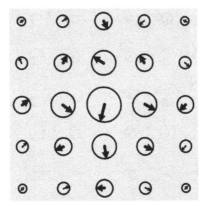

FIGURE A3.8   Wave function obtained by rotating by different amounts the dials in Figure A3.7. This change is known as a local phase transformation.

transformation of Aᵘ, the photon field, such that the fundamental equations of QED (the theory describing how electrons and positrons interact with photons) are unchanged. Note that this is not a change of phase of Aᵘ, but rather corresponds to the choices provided by classical gauge invariance, discussed above. This depends on the exact form of the interaction between electrons and photons. Another way of saying this is that by insisting on this local gauge invariance, you can predict the exact form of the electron-photon interaction. Of course, historically the interaction form came from classical electrodynamics. On the other hand, the idea that you could use gauge invariance to predict the form of interactions became the inspiration for the standard model.

Appendix 4: Unitary Matrices

A matrix is a collection of numbers organized in a table. In a unitary matrix these numbers may be of a special kind: complex numbers. Also, in a unitary matrix the numbers cannot be chosen completely at will. Once a few numbers in the table are chosen, all others are immediately set. So, although the numbers we collect in a unitary table are just that—numbers—we cannot treat them individually. Knowing a few of them completely determines the others. Unitary matrices are ubiquitous in particle physics. They appear as local symmetries in the gauge symmetries of the standard model, and they appear in the Cabibbo–Kobayashi–Maskawa mixing matrix determining CP violation within the standard model.

A4.1 A MATRIX IS A TABLE

In this book we only use square matrices. These are simply tables with as many rows as columns, as in Table A4.1.

We will use the following notation. The symbol V_{ij} refers to the number placed on the ith row of the jth column. Thus, the four symbols in Table A4.2 identify the four numbers in Table A4.1. Returning to Table A4.1, notice the following features. All numbers are smaller than 1. If you square the numbers on the first row and then sum the results, you obtain the number 1. The same happens for the second row. We say that the rows are normalized. If you sum the squares of the numbers on the first column, you get 1. The same happens for the second column. We say that the columns are normalized. If you multiply V_{11} by V_{21}, then multiply V_{12} by V_{22}, and then sum the results, you get 0. We say that the rows are orthogonal. A similar operation over columns also yields 0; the columns

TABLE A4.1   Very Simple
Example of a Unitary Matrix

0.8	0.6
−0.6	0.8

TABLE A4.2   Symbolic Representation
of a Unitary Matrix

V_{11}	V_{12}
V_{21}	V_{22}

are also orthogonal. Unitary matrices can be more complicated than the table shown here, in that they involve complex numbers.

A4.2 UNITARY MATRIX WITH FOUR ENTRIES: U(2) AND SU(2)

The dials (complex numbers) we discussed in Appendix 3 are ubiquitous in physics. In certain cases, combinations of four of these are organized into a table with four entries, such as in Figure A4.1. The table in Figure A4.1 belongs to a very special class of tables known as unitary matrices. The entries of these are complex numbers that obey the following:

$$|V_{11}|^2 + |V_{12}|^2 = 1$$

$$|V_{21}|^2 + |V_{22}|^2 = 1$$

These relations are known as the normalization of the first row and of the second row, respectively. The first relation states the following: multiply the size of each arrow of the first row by itself; sum the two results; you must get 1. The second relation says that the same operation performed on the arrows of the second row also gives 1.

The dials in special unitary matrices also obey the following:

$$|V_{11}|^2 + |V_{21}|^2 = 1$$

$$|V_{12}|^2 + |V_{22}|^2 = 1$$

which are known as the normalization of the first column and of the second column, respectively. They have the same meaning as above, but now relate entries on the same column. Normalization of rows and columns is clearly satisfied by the table in Figure A4.1. Indeed, all normalizations, on each column and on each row, involve $(0, 6)^2 + (0, 8)^2 = 1$.

Finally, unitary matrices obey the following:

$$V_{11} V^*_{21} + V_{12} V^*_{22} = 0$$

$$V_{11} V^*_{12} + V_{21} V^*_{22} = 0$$

which are known, respectively, as the orthogonality of the two rows and the orthogonality of the two columns. The first relation may be interpreted in the following way. Multiplying V_{11} with V^*_{21} gives an arrow. Multiplying V_{12} with V^*_{22} gives another arrow. Summing the two must give zero. That means that the composite arrow $V_{11} V^*_{21}$ is the negative of the composite arrow $V_{12} V^*_{22}$. Let us verify this property in Figure A4.1. Recalling that the * means that we perform top–bottom reflection, Figure A4.2a represents $V_{11} V^*_{21}$ and Figure A4.2b shows $V_{12} V^*_{22}$. We see that the results are the negative of each other, meaning that they add up to zero as required.

The set of all matrices satisfying the requirements mentioned here is known as the U(2) group, where the 2 refers to the fact that there are two columns and two rows in each matrix. These requirements relate the four entries in a very tight fashion. It turns out that if you specify the magnitude of only one entry of a U(2) table, all other sizes are immediately fixed. Also, rather than four, there are only three independent phases. In total, we only need four pieces of information to describe a U(2) table: one size and three phases.

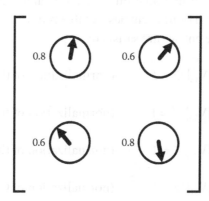

FIGURE A4.1 Table with four dials. The dial in the first row, first column will be designated by V_{11}. The dial in the first row, second column will be designated by V_{12}. The dials in the second row will be indicated by V_{21} and V_{22}, respectively.

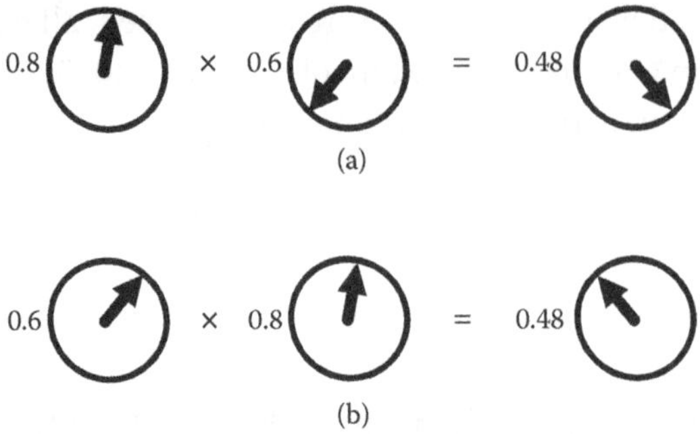

(a)

(b)

FIGURE A4.2   Dials obtained from the four dials of Figure A4.1 by performing the product: (a) $V_{11} V^*_{21}$ and (b) $V_{12} V^*_{22}$.

Certain U(2) matrices obey, in addition to all normalization and orthogonality constraints described above, the following further constraint:

$$V_{11} V_{22} - V_{12} V_{21} = 1$$

Such matrices are known as special unitary matrices and form the group SU(2). This extra condition eliminates one further independent phase. As a result, we only need three pieces of information to describe a SU(2) table: one size and two phases.

A4.3 UNITARY MATRIX WITH NINE ENTRIES: U(3) AND SU(3)

It does not take much imagination to make a table with three rows and three columns, a total of nine entries. A table is in U(3) if all the following relations among the entries are satisfied:

$|V_{11}|$^2 + $|V_{12}|$^2 + $|V_{13}|$^2 = 1   (normalization of the first row)

$|V_{21}|$^2 + $|V_{22}|$^2 + $|V_{23}|$^2 = 1   (normalization of the second row)

$|V_{31}|$^2 + $|V_{32}|$^2 + $|V_{33}|$^2 = 1   (normalization of the third row)

$|V_{11}|$^2 + $|V_{21}|$^2 + $|V_{31}|$^2 = 1   (normalization of the first column)

$|V_{12}|$^2 + $|V_{22}|$^2 + $|V_{32}|$^2 = 1   (normalization of the second column)

$|V_{13}|^2 + |V_{23}|^2 + |V_{33}|^2 = 1$ (normalization of the third column)

$V_{11} V^*_{21} + V_{12} V^*_{22} + V_{13} V^*_{23} = 0$ (orthogonality of the first and second rows)

$V_{11} V^*_{31} + V_{12} V^*_{32} + V_{13} V^*_{33} = 0$ (orthogonality of the first and third rows)

$V_{21} V^*_{31} + V_{22} V^*_{32} + V_{23} V^*_{33} = 0$ (orthogonality of the second and third rows)

$V_{11} V^*_{12} + V_{21} V^*_{22} + V_{31} V^*_{32} = 0$ (orthogonality of the first and second columns)

$V_{11} V^*_{13} + V_{21} V^*_{23} + V_{31} V^*_{33} = 0$ (orthogonality of the first and third columns)

$V_{12} V^*_{13} + V_{22} V^*_{23} + V_{32} V^*_{33} = 0$ (orthogonality of the second and third columns)

Written in this way, the relations look very daunting. But they are easy to remember with the aid of some graphical mnemonics. For example, consider the matrix in Figure A4.3. We can represent all the normalizations

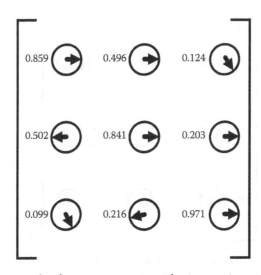

FIGURE A4.3   Example of a unitary matrix with nine entries.

of rows and columns by Figure A4.4. For example, for the normalization along the first row, we do the following. We calculate the square of the magnitude of all elements along the first row; we add those numbers together; the result must equal 1. Remember that the magnitude of a dial gives the number; i.e., the phase is irrelevant for the magnitude squared. The orthogonality of the second and third rows is represented in Figure A4.5. This represents the following set of operations: find the conjugate of all

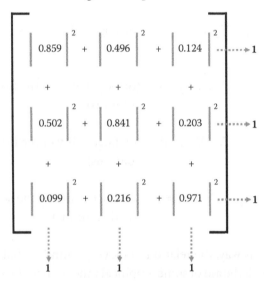

FIGURE A4.4 Normalization of rows and columns of a U(3) table of complex numbers.

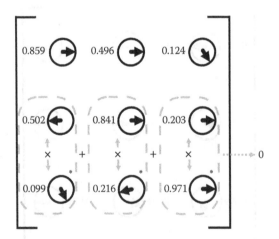

FIGURE A4.5 Orthogonality between the second and third rows of a U(3) table of complex numbers.

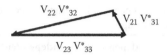

FIGURE A4.6   Triangle representing the orthogonality between the second and third rows of the matrix in Figure A4.5.

elements along the third row; multiply each element on the second row by the (conjugate of the) corresponding element on the third row; adding all results together, you must get 0.

An interesting graphical representation of this orthogonality relation is the following. The first dashed grey box corresponds to $V_{21} V^*_{31}$, the second to $V_{22} V^*_{32}$, and the third to $V_{23} V^*_{33}$. Each is a complex number, and thus can be viewed as an arrow. Stating that the three arrows must add to zero is the same as stating that these must form a triangle, as shown in Figure A4.6. Similar graphical representations hold for the orthogonalities between different pairs of rows or between pairs of columns. They are known as the unitarity triangles.

Again, the relations shown place a severe constraint on the nine entries in the table. It turns out that if you specify the magnitudes of only three entries of a U(3) table, all other magnitudes are immediately fixed. Also, rather than nine, there are only six independent phases. The set of all U(3) tables with an appropriate multiplication rule (which we do not discuss) forms a group. In total, we only need nine pieces of information to describe a U(3) table: three sizes and six phases.

Certain U(3) matrices obey an extra complicated condition involving all entries. These are known as special unitary matrices and form the group SU(3). This relation eliminates one further independent phase. As a result, we only need eight pieces of information to describe an SU(3) table: three sizes and five phases.

A4.4 INDEPENDENT PARAMETERS IN UNITARY MATRICES AND THE NUMBER OF GAUGE BOSONS

The information needed to describe completely the various unitary matrices we have discussed is listed in Table A4.3.

The standard model is based on the gauge groups SU(3), SU(2), and U(1). Applying the gauge principle to these groups means that a messenger particle is needed for each independent piece of information required in order to describe a group element. Thus, we need eight gluons to carry the SU(3)-based strong interaction and 3 + 1 = 4 gauge bosons (three weak

TABLE A4.3   Number of Independent Parameters Needed in Order to Describe
an Element of the Unitary Groups

Group	Number of Entries	Independent Magnitudes (sizes)	Independent Phases (angles)	Number of Information Pieces Required
U(1)	1	0 (dial has size 1)	1	1
SU(2)	4	1	2	3
U(2)	4	1	3	4
SU(3)	9	3	5	8
U(3)	9	3	6	9

gauge bosons and the photon) to describe the SU(2) × U(1)-based electroweak interactions.

A4.5 THE CABIBBO–KOBAYASHI–MASKAWA MATRIX

A unitary matrix is used to relate one set of basic states to another. A very
important example is the Cabibbo–Kobayashi–Maskawa matrix introduced in Chapter 9. There, after introducing charm, we replaced the set (d,
s) by the set (d_W, s_W); see Figure 9.4. The unitary matrix relates the one set
to the other. This is completely analogous to rotation of coordinates going
from (x, y) to (x′, y′), as illustrated in Figure A4.7. Given the arrows (A and
B) of unit length along the x and y axes, we find that they are given in the
(x′, y′) reference frame by the four entries in the matrix of Table A4.1. Note
that the sum of the squares of each row and each column equal 1 since the
rotation does not change the length of the arrows. Also there is the orthogonality relation (0.8) (−0.6) + (0.6) (0.8) = 0, which expresses the fact that
the arrows are perpendicular to each other. These are the relations that
define a unitary matrix. Given these relations the matrix is determined by
a single parameter, the angle of rotation theta.

For the case of the CKM matrix we go from the set (d, s, b) to (d_W, s_W, b_W),
which requires a 3 × 3 unitary matrix. Similarly, when we come to neutri-
nos we go from the weak states (v_e, v_μ, v_τ) to the mass states (v_1, v_2, v_3). The
same conditions concerning the sum of the squares and the orthogonality
relations hold. These conditions have the consequence that the matrix is
determined by three parameters. One could envision this as a rotation in
three-dimensional space where there would be three axes of rotation.

As noted before, fields are represented by complex quantities so that in
general the unitary matrices may contain complex numbers. In the most
general case, one gets three independent phases for a two-dimensional

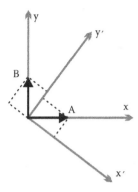

FIGURE A4.7 Description of the vectors A and B in two coordinate systems (x, y) and (x′, y′), rotated with respect to each other.

unitary matrix, as shown in Table A4.3. However, these phases have no physical consequence in the case where the unitary matrix describes the change from the set (d, s) to the set (d_W, s_W). The reason is as follows. Dirac tells us that we can alter the phase of the field corresponding to the state d, without any physical consequence (see Appendix 3). Although changing all states (d, s, d_W, and s_W) by the same phase has no effect at all, we still have three relative phases among these four states. These can be used to absorb the three phases in the unitary mixing matrix. This has a very important consequence. A theory with only four quarks (u, d, s, and c) does not exhibit CP violation.

Similarly, Table A4.3 indicates that a general three-dimensional unitary matrix has six independent phases. However, when this matrix describes the change from the set (d, s, b) to (d_W, s_W, b_W), we can remove five relative phases from the problem. Thus, the Cabibbo–Kobayashi–Maskawa matrix describing that mixing has only one physically relevant phase. As pointed out in Part C, this phase is the only source of CP violation in the standard model. This mechanism for CP violation, originally proposed by Kobayashi and Maskawa, earned them the 2008 Nobel Prize in Physics, after it was verified experimentally at the B factories in the early twenty-first century.

Appendix 5: Energy and Momentum in Special Relativity and the Uncertainty Principle

In special relativity, the energy, E, of a particle is given by

$$E^2 = c^2 p^2 + m^2 c^4 \qquad (A5.1)$$

where p is the momentum, m is the mass, and c is the velocity of light. We call this the energy–momentum–mass relation. For a particle at rest, $p = 0$, and we obtain the famous Einstein equation

$$E = mc^2$$

which loosely states that mass is just another form of energy. For a massive particle moving slowly, Equation A5.1 gives to a good approximation

$$E = mc^2 + \frac{p^2}{2m} \qquad (A5.2)$$

To check that Equation A5.2 is a good approximation of Equation A5.1, you may square Equation A5.2 and neglect the small term of order p^4. Writing $p = mv$ (where v is the particle's velocity), the second term in Equation A5.2 becomes $\frac{1}{2} mv^2$, which is the form of the kinetic energy usually seen in high school physics.

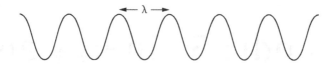

FIGURE A5.1   Snapshot of a wave at a given time. The wave cannot be drawn here in its entirety; it starts at the infinitely far left and ends at the infinitely far right. The distance between two consecutive crests is known as the wavelength and represented by the Greek letter λ (lambda).

For massless particles like the photon, $m = 0$, and Equation A5.1 gives

$$E = c\,p \tag{A5.3}$$

This relation between energy and momentum holds in classical physics for electromagnetic waves.* For the case of the neutrino, which has a very small mass, we have to good approximation

$$E = c\,p + \frac{m^2 c^4}{2 c\,p} \tag{A5.4}$$

As above, you can check that Equation A5.4 agrees with Equation A5.1 if one neglects terms of order m^4.

Considering neutrino oscillations, we are interested in the energy difference between two neutrinos, v_1 and v_2, with a given momentum p. From Equation A5.4,[†]

$$E_2 - E_1 = \frac{\left(m_2^2 - m_1^2\right)c^4}{2 c\,p} \tag{A5.5}$$

Treating the neutrino as a wave, its frequency is $f = E/h$, where h is Planck's constant. The phase difference between v_2 and v_1 over time is determined by

$$\left(E_2 - E_1\right)\frac{t}{h} = \frac{\left(m_2^2 - m_1^2\right)c^4}{2 c\,p}\frac{t}{h} \tag{A5.6}$$

In Figure 16.4, we illustrate the phase difference as a function of time. The neutrino starts out as the combination $(v_2 + v_1)$. When the time reaches

---

* In fact, classical electromagnetism was the inspiration for special relativity.
† Since the masses are very small the p in the denominator may be replaced by the energy E, which is the average energy of the two neutrinos.

FIGURE A5.2   Snapshot of a finite wave at a given time.

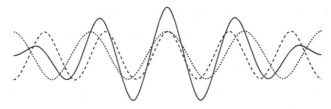

FIGURE A5.3   Superposition of two waves of slightly different wavelengths. The two waves are shown as grey (solid and dashed) lines; their sum is shown as a solid black line.

t_1 the state is $(v_2 - v_1)$. At the later time, t_2, the state is $(v_2 + v_1)$ again. Thus, ct_2 is the oscillation wavelength.

There is a fundamental problem about wave particles that still needs to be addressed. A wave with definite momentum p has a wavelength h/p, which describes an endless wave. Consider a wave moving to the right. Taking a snapshot of the wave at a given time we see Figure A5.1. The amplitude of the wave as a function of x is given by

$$\psi = A \sin(2\pi x / \lambda) \qquad (A5.7)$$

where λ is the wavelength. This function goes on forever; it describes an ideal wave with absolutely well-defined wavelength. It has no beginning and no end. Real waves have a finite extent. You ring the bell, the wave starts; the bell stops ringing, the wave stops. The amplitude of such a finite wave (in one direction, along the x axis) looks like Figure A5.2. A wave of finite extent is represented by a so-called wave packet. The way to construct a wave packet is to combine a set of ideal waves, given by Equation A5.7, each with a slightly different wavelength and amplitude, so that the set of wavelengths covers the range λ_0 to $\lambda_0 + \Delta\lambda$. The idea is illustrated in Figure A5.3, where two waves with slightly different wavelengths are superimposed. The solid and dashed grey lines add to the black line. Near the center the two waves interfere constructively. In contrast, at the edges of the figure, the two waves interfere destructively and cancel each other out. The smaller the range of wavelengths

$\Delta\lambda$, the longer is the length $\overline{X}$ of the wave packet before the destructive interference takes over.

If there were only two wavelengths in the picture (as in Figure A5.2), there would be constructive interference again after a while. This phenomenon is known as beats and is familiar from music when two nearly equal notes are struck simultaneously. However, if you combine all the wavelengths in the interval λ_0 to $\lambda_0 + \Delta\lambda$, there is destructive interference everywhere outside the length $\overline{X}$.

How long is $\overline{X}$? In order to get destructive interference, one must go a distance x in either direction such that

$$\frac{2\pi x}{\lambda+\Delta\lambda} = \frac{2\pi x}{\lambda} - \pi \tag{A5.8}$$

If $\Delta\lambda$ is much smaller than λ, Equation A5.8 can be approximated by

$$x = \frac{\lambda^2}{2\Delta\lambda} \tag{A5.9}$$

Since this occurs in both directions, we multiply by 2, and estimate the order of magnitude of $\overline{X}$:

$$\overline{X} = \frac{\lambda^2}{\Delta\lambda} = \lambda\left(\frac{\lambda}{\Delta\lambda}\right) \tag{A5.10}$$

One way to think about this equation is the following. Each time you go through one cycle (x increases by λ so that $2\pi x/\lambda$ increases by 2π), you get the two waves in the picture out of phase by a fraction $\Delta\lambda/\lambda$. Thus, they are totally out of phase for n cycles when $n = \lambda/\Delta\lambda$. This corresponds to a change in x of order $n\lambda = \lambda^2/\Delta\lambda$.

To summarize, a wave of finite spatial extent may be considered as the superposition of waves of different wavelengths. Assuming the waves are all moving to the right, then the whole wave packet of finite extent is moving to the right. The shorter the length of the wave packet, the larger the range of the wavelengths that must be superimposed.

Given a wave particle of momentum p_0, its wavelength is given by $\lambda = h/p_0$, and its amplitude is given by

$$\psi = A \sin(2\pi p_0 x/h) \qquad \text{(A5.11)}$$

This is a wave of infinite extent. To localize the particle so that it is known to be inside a region of extent $\overline{X}$, we have to construct a wave packet with different wavelengths superimposed. This means different momenta. Since λ and p are linearly related, a small percentage change in p produces a small percentage change in λ, related by

$$\frac{\Delta\lambda}{\lambda} = \frac{\Delta p}{p} \qquad \text{(A5.12)}$$

Substituting this in Equation A5.10 gives

$$\overline{X} = \lambda \left(\frac{p}{\Delta p} \right) \qquad \text{(A5.13)}$$

Using $\lambda = h/p$ and letting $\overline{X} = \Delta x$, we find

$$\Delta x \, \Delta p = h \qquad \text{(A5.14)}$$

This is the famous Heisenberg uncertainty principle. This is an order magnitude relation. The important point is that the more precisely you try to pin down the position, the more uncertain is the momentum. The uncertainty principle in Equation A5.14 is usually given as $\Delta x \, \Delta p \geq h/2\pi$.

Notice that Equation A5.14 results from the quantum mechanical prescription that a particle has an associated wave of probability. This prescription is the crucial step. Once it is made, the quantum mechanical wave particle must obey Equation A5.14.

If we think of a particle (a photon or electron, for example) being represented by a wave packet, then we cannot predict exactly the position or the momentum of the particle. There is a probability distribution for the position and another probability distribution for the momentum. Figure A5.4 shows a curve $Q(x)$ for the probability that a measurement of the particle's position will give a value between x and x + Δx, with a similar curve $P(p)$ for the momentum. The width of the curve $Q(x)$ is about Δx, and that of $P(p)$ is about Δp, such that Equation A5.14 is satisfied.

In analyzing neutrino oscillations, we should think of the neutrinos as produced in a wave packet rather than a precise momentum. However, in practice this does not change the results of the simplified analysis we have presented.

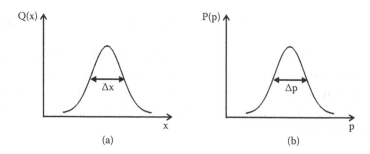

FIGURE A5.4 (a) Curve Q(x) for the probability that a measurement of the particle's position will give a value between x and x + Δx. (b) Similar curve P(p) for the momentum.

Instead of considering the wave as it varies in space, we can look at it as a function of time with the frequency, f, given by E/h, where E is the energy. An argument similar to that given above yields another uncertainty relation:

$$\Delta E\, \Delta t \geq h/2\pi \qquad (A5.15)$$

Observations limited to a time interval Δt can only determine the energy with an uncertainty ΔE.

One application of this is to a decaying particle. The particle exists only for a limited time measured by the decay lifetime, τ. Thus, there is an uncertainty in the energy. Assuming the decaying particle is at rest, this means an uncertainty in the mass because $E = mc^2$. Thus, applying Equation A5.15,

$$\Delta\left(mc^2\right) \geq h/2\pi\tau \qquad (A5.16)$$

The value of $\Delta(mc^2)$ is called the width of the particle decay. In the example of the psi particle discussed in Chapter 6 we see in Figure 6.5 that the energy needed to produce the J/psi covers an interval. While we call the peak of the curve the mass of the J/psi, the spread around the peak illustrates the width of the decay, which could be considered uncertainty in the mass. In this case the width is rather large because the J/psi has a very short lifetime, less than a billionth of a billionth of a second.

It was first pointed out by Yukawa (as discussed in Chapter 5) that the range R of an interaction depends on the mass M of the virtual particle exchanged between the interacting particles. The larger the mass M, the shorter the range R, that is, the smaller the distance over which the interaction is effective. More precisely, the interaction falls off exponentially with distance so that it is strongest at distances less than R and is very weak at distances much larger than R. Gian-Carlo Wick provided a qualitative understanding of the range R using the uncertainty relation of Equation A5.15. Considering the interacting particle at rest, the creation of the virtual particle with mass M violates energy conservation by an amount $\Delta E = Mc^2$. It is then argued that this can only happen over a time Δt of the order h/Mc^2. During the time Δt the particle with mass M could travel at most a distance R given by

$$R = c\,\Delta t = h/Mc \qquad\qquad (A5.17)$$

This is the range originally given in Yukawa's theory. The pion has a mass of about 140 MeV (1/7 of the mass of the proton) so that R is about 10^{-13} cm. For the case of the weak interaction, the virtual particle exchanged is the W boson with a mass of 80 GeV (almost ninety times that for the proton), so that the range is of the order 10^{-16} cm. In Fermi's original theory of the weak interaction he assumed the interaction occurred at a point, essentially $R = 0$. For low-energy processes like beta decay, this is a perfectly adequate approximation. The mass of the W boson only shows up in high-energy processes.

Index

F
_____

G
_____